Global Energy Politics

Global Energy Politics

Thijs Van de Graaf
Benjamin K. Sovacool

polity

First published in 2020 by Polity Press

Reprinted 2020

Polity Press
65 Bridge Street
Cambridge CB2 1UR, UK

Polity Press
101 Station Landing
Suite 300
Medford, MA 02155, USA

ISBN-13: 978-1-5095-3048-9
ISBN-13: 978-1-5095-3049-6(pb)

A catalogue record for this book is available from the British Library.

Library of Congress Cataloging-in-Publication Data
Names: Graaf, Thijs van de, author. | Sovacool, Benjamin K., author.
Title: Global energy politics / Thijs Van de Graaf and Benjamin K. Sovacool.
Description: Cambridge, UK ; Medford, MA : Polity, 2020. | Includes bibliographical references and index. | Summary: "Van De Graaf and Sovacool uncover the intricate ways in which our energy systems have shaped global outcomes in areas of world politics"-- Provided by publisher.
Identifiers: LCCN 2019038076 (print) | LCCN 2019038077 (ebook) | ISBN 9781509530489 (hardback) | ISBN 9781509530496 (paperback) | ISBN 9781509530519 (epub)
Subjects: LCSH: Energy policy. | Energy development--Political aspects. | Energy conservation--Political aspects. | Energy industries--Political aspects.
Classification: LCC HD9502.A2 G678 2020 (print) | LCC HD9502.A2 (ebook) | DDC 333.79--dc23
LC record available at https://lccn.loc.gov/2019038076
LC ebook record available at https://lccn.loc.gov/2019038077

Typeset in 10.5 on 12 pt Plantin by
Servis Filmsetting Ltd, Stockport, Cheshire
Printed and bound in Great Britain by TJ Books Limited

For further information on Polity, visit our website: politybooks.com

Contents

Figures, Tables, and Boxes

Figures

Tables

Boxes

Foreword

We are living in an age of unprecedented global change, which is affecting all facets of our societies. Digital technologies are disrupting existing economic and social structures, established political and trade systems are under pressure, and extreme weather conditions are a regular reminder of the perils of climate change for our planet.

Similar forces are also reshaping the energy sector. As Director-General of the International Renewable Energy Agency (IRENA) from 2011 to 2019, I have had the privilege to witness first-hand this ongoing energy transformation. In less than a decade, renewable energy has moved from the fringes to the center stage of the global energy landscape, thanks to supportive policy frameworks, technological innovation, and rapidly decreasing costs. In 2018, a record-breaking 171 GW of renewable energy capacity were added globally, led by emerging and developing economies, making it the seventh consecutive year in which additional power generation capacity from renewables outpaced conventional sources. A third of global power capacity is now based on renewables.

Few would have envisioned such remarkable progress just some years ago. We are truly in the midst of revolutionary transformations in our global energy system. The change is driven by the compelling business case for renewables. According to IRENA's analysis, in 2020, all currently available mainstream renewable energy technologies will be cost-competitive with fossil fuels in most parts of the world. This is a momentous change, especially if you consider that countries taking full advantage of their renewables' potential will also benefit from a host of socioeconomic benefits, including lower carbon emissions, cleaner air, and more sustainable job growth.

Renewable energy deployment needs to grow even faster, however, to ensure that we can achieve the global climate objectives and the Sustainable Development Goals. The latest special report of the International Panel on Climate Change on Global Warming of 1.5°C has shown that half a degree makes a world of difference and that there is really only one temperature goal toward which we should orient ourselves. Exceeding the ceiling of 1.5°C would lead to intolerable loss of biodiversity, destruction of infrastructure, and many more people exposed to life-threatening conditions such as extreme heat. IRENA's analysis has shown that renewables and energy efficiency combined provide the most cost-effective pathway to achieve 90% of energy-related reductions required to meet the well below 2°C objective of the Paris Agreement.

As someone who comes from a developing country, Kenya, the plight of energy poverty, which affects billions of people, is very dear to my heart. I believe that the energy transformation offers an opportunity to shift the global development paradigm from one of scarcity, inequality, and competition to one of shared prosperity – in our lifetimes. That is an opportunity we must rally behind by adopting strong policies, mobilizing capital, and driving innovation across the energy system.

In my capacity as head of IRENA, it became clear to me that the energy transformation involves more than the transition from one energy source to another. It also entails the transformation of the geopolitical landscape with profound implications for societies and economies. It is about giving jobs to the millions of young people in emerging and developing economies, generating prosperity at large in an age of austerity, and the reconfiguration of international relations. That is why, in 2018, I took the initiative to set up the Global Commission on the Geopolitics of Energy Transformation as an independent body to look further into the wider ramifications of the energy shift for global wealth and stability.

Its final report, which was presented to IRENA's 160 member states at the General Assembly in January 2019, is a first foray to chart the new geopolitical world emerging from the energy transformation driven by renewables. It lays out how the energy transformation will reshape relations between states and will change the global distribution of power. As countries increasingly develop renewables at home and integrate their grids with those of neighboring countries, they will achieve greater energy independence. Fossil-fuel importing countries will improve their trade balance and enjoy significant macroeconomic benefits. Countries that lead in clean technology innovation stand to gain from the global energy transformation. There will be a diffusion of power and new actors will become more prominent. As renewables decentralize and democratize energy systems, citizens, cities, and regions will become increasingly important players in the new energy landscape.

Certainly, the energy transformation brings new risks related to cybersecurity, new dependencies on critical materials, and socio-economic dislocation in certain countries and sectors. The energy transition might be more of a bumpy ride than a smooth sailing. But, as the report found, overall the benefits of the energy transformation will outweigh the challenges, provided that the right policies and strategies are developed to mitigate the risks. This applies particularly to fossil-fuel exporters. The need for them to diversify their economy offers a huge opportunity to enhance long-term growth, create new jobs, and bring about a more productive future.

In a complex and rapidly changing energy landscape, a comprehensive overview of the global politics of energy such as this one is particularly welcome. I am delighted to recommend this insightful book by Thijs Van de Graaf and Benjamin Sovacool, who skillfully connect the dots between energy markets, geopolitics, the environment, and local activism across a range of energy technologies and sectors. For anyone who wants to understand the complexities and depth of the global energy challenge, *Global Energy Politics* is essential reading.

Adnan Z. Amin, former Director-General of
the International Renewable Energy Agency (IRENA)

Preface

As we are finishing this book manuscript, in the summer of 2019, energy and climate issues are again capturing global headlines. Oil tankers are being seized in the Strait of Hormuz, where tensions have been rising since the United States withdrew from the nuclear deal with Iran. A heatwave that shattered records in Europe moved to Greenland, where it triggered massive ice melt as well as forest fires. For the first time since the Industrial Revolution, the UK is reported to generate more electricity from zero-carbon sources than from fossil fuels. The Welsh and UK Parliaments also declared a "climate emergency" over the summer. The United States Congress is taking steps to adopt a Bill that sanctions companies that help to build new gas pipelines from Russia to Europe. These few examples show how the energy and climate conundrum is one of the key undercurrents of world politics in the twenty-first century.

This book aims to introduce readers to the global politics of energy, at a time of momentous changes in both the energy system and global geopolitics. Before embarking on our journey, we would like to thank a number of people. At Polity, we are grateful for the trust and support of Louise Knight, Inès Boxman, and Sophie Wright. We thank Mathieu Blondeel and Moniek de Jong for their assistance with several sections and graphs, as well as Kingsmill Bond for commenting on an earlier version of this manuscript. Both of the authors thank their wonderful families for their continued support, and incredible patience. Last but not least, Thijs Van de Graaf wishes to commend the past and present students that follow his course 'Global Energy Politics'. These students' curiosity, questions and critical reflections have made enormous contributions to his thinking on the subject, and have helped to shape this book from its early conception to its final form.

Thijs Van de Graaf & Benjamin K. Sovacool
August 2019

About the Authors

Thijs Van de Graaf is Associate Professor of International Politics at Ghent University, Belgium. He is also a non-resident fellow with the Payne Institute, Colorado School of Mines, and with the Initiative for Sustainable Energy Policy (ISEP) at Johns Hopkins University. He was the lead drafter of the report '*A New World: The Geopolitics of the Energy Transformation*' (2019), commissioned by the International Renewable Energy Agency (IRENA). His recent books include *The Palgrave Handbook of the International Political Economy of Energy* (Palgrave, 2016), *The Politics and Institutions of Global Energy Governance* (Palgrave, 2013), and *Global Energy Governance in a Multipolar World* (Ashgate/Routledge, 2010).

Benjamin K. Sovacool is Professor of Energy Policy at the Science Policy Research Unit (SPRU) at the University of Sussex Business School, United Kingdom. He is also Director of the Center for Energy Technologies and Professor of Business and Social Sciences in the Department of Business Development and Technology at Aarhus University, Denmark. He is a Lead Author of the Intergovernmental Panel on Climate Change's Sixth Assessment Report (AR6), due to be published in 2022, and an Advisor on Energy to the European Commission's Directorate General for Research and Innovation in Brussels, Belgium. With much coverage of his work in the international news media, he is one of the most highly cited global researchers on issues bearing on controversies in energy and climate policy.

1

Introduction: Systems, frames, and transitions

Energy is central to almost all areas of modern human activity. The computer on which we are typing this text, the smartphone in your pocket, and the heating in your home all depend on the availability of sufficient and affordable supplies of energy.[1] Access to energy services is often taken for granted in affluent countries. Yet, almost a billion people in the world do not have access to electricity in their homes, while many more suffer from supply that is of poor quality.[2] These people experience on a daily basis what it means to have no or insufficient access to mobility, lighting, refrigeration, or telecommunication. In very simple terms, many of these people have never watched a television show, let alone streamed Netflix; they have never had a hot shower, let alone stayed in a modern hotel; they may never have had a cold beer, let alone refrigeration for vaccines. Moreover, energy is the single largest source of greenhouse gas (GHG) emissions, resulting in climate change that, if left unchecked, could devastate our planet.

Energy does not just affect our personal daily lives and our natural ecosystems. It also profoundly shapes our wider societies, economies, and politics – and it has done so throughout history, from pre-agricultural foraging societies through today's fossil fuel-driven civilization.[3] Few people realize the extent to which the wealth of nations, the fate of political leaders, and international relations more broadly are shaped by the way we produce and consume energy. Energy is not just another commodity: it is a strategic good for the survival of regimes, a critical input factor for the world economy that can shift large swaths of wealth around, a massive source of pollution, and a major cause of social goods and evils. These attributes make energy a key driver of the pursuit of wealth and power in world politics.

The upshot is that energy questions are deeply political and yield distributional consequences. They create winners and losers. That is why the global energy challenge of ensuring secure, sustainable and affordable access to energy for all is not susceptible to resolution by the "hard" or "objective" disciplines of physics, mathematics, economics, and engineering. Such an approach fails to grasp the political stakes and tradeoffs involved in energy trade and decision-making, from the global all the way to the local and even individual level. Energy discussions are often reduced to technical issues and matters of cost but, at their core, they involve political and moral choices about the kind of society we want to live in.

Even so, scholars of international relations (IR) and social scientists more generally have long overlooked energy issues.[4] To the extent that global energy politics is addressed at all by IR scholars, it is mostly framed through geopolitical lenses, and it focuses almost exclusively on oil and gas.[5] The concept of "energy security," in particular, has become something of a cottage industry, spawning a voluminous body of work.[6] Recently, a small number of studies have sought to counter this security focus by adopting perspectives rooted in public policy,[7] governance,[8] and international political economy.[9] Unfortunately, however, there has been little effort to integrate these various dimensions into a single framework.

The goal of this book is to provide an overview of the main concepts and approaches in the study of "global energy politics," which we define as the struggle over who gets what, when, and how in energy use and production from an international perspective. The book introduces a novel framework to interpret global energy politics, based on a socio-technical system approach and contested frames. *Socio-technical systems* refers to a broad conceptualization of the infrastructures in place to deliver energy services, not just as the resources and technologies themselves, but also as a set of user practices, cultural meanings, institutions, and supply networks. It goes much beyond narrow views of energy security as the secure supplies of oil and natural gas, and it illustrates how the politics of coal is different than that of oil or renewable energy. *Contested frames* refers to different stakeholder views surrounding different energy systems: neo-mercantilism, market liberalism, environmentalism, and egalitarianism. Each reflects different ideologies, world views, value-systems, and hard-nosed interests.

Key questions that we will address include: How can we ensure that all people have reliable and affordable access to sufficient energy for their needs? When and how will states cooperate to mitigate

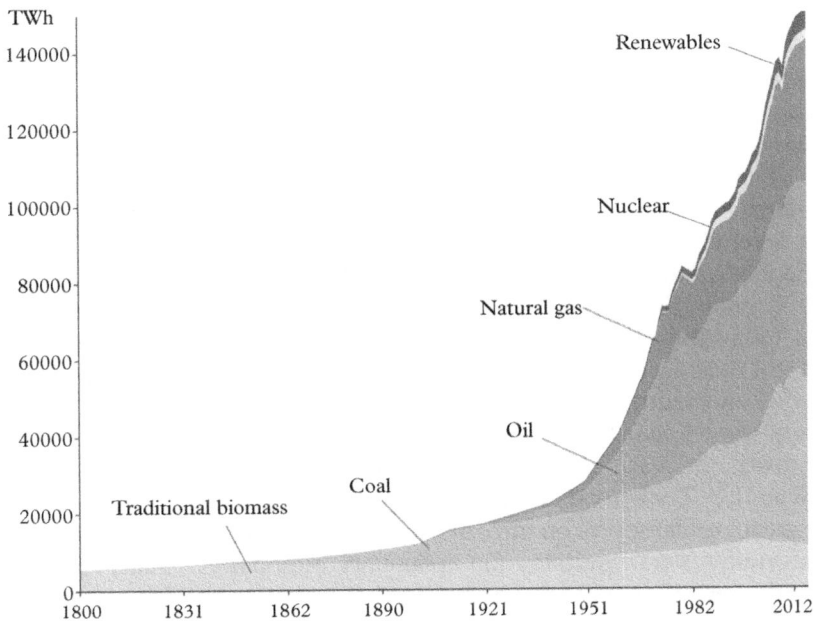

Figure 1.1 Evolution of the modern energy system

Source: Compiled with data from Smil, V. (2017) *Energy Transitions: Global and National Perspectives*. Santa Barbara, CA: Praeger, Appendix A.

energy-related emissions of greenhouse gases? In what ways can energy be used as a foreign policy tool to coerce other countries? What does it mean to make a "just transition" to low-carbon energy sources and how could it affect energy democracy and local politics?

A critical juncture in energy politics

The story of the modern energy system over the past two centuries is primarily one of rapidly increasing use of fossil fuels – that is, oil, coal, and natural gas (see figure 1.1). During the Industrial Revolution in the nineteenth century, coal gradually replaced wood and biomass. Coal was itself overtaken by oil after World War II as the world's dominant energy source. Natural gas use has also increased globally, but as of now, oil still reigns supreme. Fossil fuels today still supply around 80 percent of worldwide primary energy consumption. Renewables are among the fastest-growing energy sources in recent years, but since they start from a very low base, it takes considerable time for them to make a mark on the global energy mix. It is well worth noting that the energy shifts in figure 1.1 represent energy *additions* rather than energy *transitions*. For example, even if coal has lost market share relative to oil and gas in the latter half of the twentieth century, it has still continued to grow in absolute terms.[10]

The fossil fuel era has produced great wealth as well as advancements in convenience, comfort, and cleanliness.[11] It has sustained a sevenfold increase in population growth – to more than 7 billion people – and a seventyfold increase in global production over the last 200 years. By this measure, the "average" inhabitant of planet earth is today more than 11 times better off than in 1820.[12] Exploiting fossil fuels has set off an energy bonanza, transforming human societies and cultural values, making them in a way more democratic and less violent.[13] These trends – energy consumption, population growth, increase in wealth, and others – accelerated after World War II.[14] Today, we are also witnessing the growing imprint of this "great acceleration" on our environment, and the possibility that our economy is transgressing the "planetary boundaries" that provide a "safe operating space for humanity," threatening the functioning of ecosystems and triggering destructive climate change.[15]

Fossil fuels have several advantages which make them attractive energy sources: they have high energy densities (particularly oil), are easy to store and transport (particularly coal and oil), and they are versatile in their applications (natural gas, for instance, is used for cooking, heating, power generation, mobility, and as a chemical feedstock). They are also relatively cheap, but often that is because their full environmental and health impacts are not well reflected in their price. Fossil fuels are not just polluting, they are also stored in finite reserves that are geographically concentrated, raising energy security concerns for countries that rely on imports.

Today, we are on the cusp of four major transformations in global energy politics, transformations that will not just disrupt the way we produce and consume energy, but will also likely transform our economies, societies, and political systems along the way. The aftershocks of these energy transformations will be felt in corporate boardrooms, political cabinets, and individual households. These changes have important implications for how we study and define the global politics of energy. For decades, global energy politics has been synonymous with the struggle for oil, the power game between oil-rich countries, private oil companies, and Western consumer countries. Such a narrow focus is no longer tenable in light of the four tectonic shifts that are afoot in the global energy system.

First, *climate change*, if unfettered, is posing an existential threat to life as we know it on our planet. Numerous reports of the Intergovernmental Panel on Climate Change (IPCC) have demonstrated the urgent need to decarbonize the global economy, and they have become more alarming about the calamitous consequences for our ecosystems if we fail to heed its warnings. Climate change is blamed on many human activities, including agriculture and deforestation (which is called "land-use change" in IPCC speak), but by far the main culprit is our massive burning of fossil fuels. Fossil fuels account for 80 percent of worldwide primary energy but they also account for around 80 percent of carbon dioxide (CO_2) emissions, and the energy sector is also a key source of other greenhouse gases such as methane (CH_4). So, to fix climate change we need to change the way we produce and consume energy. Climate change is mostly an energy problem.

Second, the *rise of China* has sent shockwaves through world energy markets. The sheer size of China's energy demand, and its key position in global value chains of critical energy resources and technologies, have made the country a category in its own right. It is by far the world's largest energy consumer and greenhouse gas emitter, and it has also emerged as the biggest producer of renewable technologies such as solar panels and electric vehicles. Depending on which statistics you rely on, China is the world's largest energy consumer, the biggest emitter of greenhouse gases, fifth largest producer of oil, seventh largest producer of natural gas, and the largest miner of coal.[16] Over the past five years, over 65 million new jobs were created in the Chinese economy.[17] The country now leads the world in markets for automobiles, steel, cement, glass, housing, power plants, renewable energy, highways, rail systems, and airports.[18] The future of our energy system therefore hinges to a large extent on decisions made in Beijing. China sets the shape of the market. China represents a larger category of populous, emerging economies like India, Brazil, and Indonesia, who are becoming important engines of global energy demand growth. It is clear that, like in the world economy in general, the center of gravity in global energy politics is shifting from West to East.

Third, we are on the verge of a major *transition to renewable energy*, thanks to technological advances and dramatic falls in cost. The rapid growth of solar and wind power, in particular, over the past decade has surprised even the most optimistic industry players and observers. The shift from fossil fuels to renewables is not just a shift from one fuel to another, as there are important differences between the two. While fossil fuels are geographically concentrated stocks of energy, renewable energy sources like wind, water, and sun are ubiquitous flows of energy; they are available in one form or another in most countries and they cannot be exhausted. Moreover, renewables can be deployed at almost any scale and lend themselves better to decentralized forms of energy production and consumption. Modern renewables, finally, have nearly zero marginal costs, and some of them, like solar and wind, enjoy cost reductions of more than 20 percent for every doubling of capacity.[19] The shift to renewables thus involves a deep transformation of energy systems which is likely to affect global trade patterns, blur the distinction between producers and consumers, and create new patterns of political authority along the decentralized deployment of renewable technologies.

Fourth, the low-carbon transition will have to happen simultaneously to a major push to *eradicate energy poverty*. Large parts of the world's population, notably in Africa and India, still do not have access to electricity or clean cooking facilities. Bringing energy services to billions of households and companies in the developing world is a moral imperative that is often overlooked in standard accounts of global energy security. It is embodied in universal energy access via initiatives such as Sustainable Energy for All and the Sustainable Development Goal 7.

The scope and approach of this book

Against the backdrop of these tectonic shifts and looming challenges in the global energy system, our goal is to engage a three-way relationship between: (1) the world energy system, characterized by the four major transformations outlined above; (2) relations between countries; and (3) domestic energy politics and governance (see figure 1.2). We will thus examine how shifting patterns of energy consumption and production affect relations between countries and domestic politics in energy-producing and -consuming countries, but also explore how domestic energy politics affects relations between countries and how both international relations (IR) concerns and domestic energy politics are shaping energy production and consumption patterns.

As previously indicated, IR scholars have long overlooked energy issues. The oil crises of the 1970s generated some attention to the issue of oil security and energy markets, and the recent turmoil in

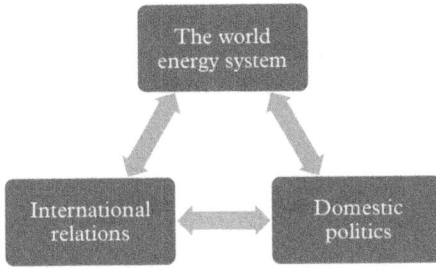

Figure 1.2   The three-way relationship between energy, international, and domestic politics

Source: Authors.

global oil markets revived that attention. However, there is a more troubling void in terms of analytical frameworks to analyze global energy politics. Much of the extant writings on energy from an IR perspective have adopted a narrow geopolitical lens that corresponds largely to realist and neo-realist thinking. They focus mostly on oil and gas, and regard these fuels as a currency for power or a source of vulnerability. States and their national energy companies are seen as the most pivotal players and international energy transactions are portrayed as zero-sum exchanges.

Our analytical approach differs from that mainstream view in two ways. First, rather than limiting ourselves to the supply of just oil and gas, we adopt a much broader, socio-technical systems perspective on energy. The energy system, in our conceptualization, comprises not just the supply infrastructure, but also the demand infrastructure and the social contexts in which these supply and demand infrastructures are embedded. We focus not just on oil and gas, or even primary resources, but also on prime movers (e.g., household appliances, cars), the built environment (e.g., insulation of buildings), and the political and cultural elements that shape our energy behavior. This constellation of energy systems is what makes "energy" distinct from other political spheres (e.g., "global food politics" or "international relations" in general). It even makes it "unique" compared to other pressing global governance challenges such as health or trade.[20]

Second, we argue that this energy system, broadly conceived, can be seen through four archetypical frames. Each frame – a guiding worldview – comes with a different problem definition, diagnosis of the causes, and suggested remedy. The mainstream, realist-inspired view of oil and gas trade as zero-sum relations corresponds to the frame of neo-mercantilism. This frame can be complemented with three others: market liberalism, environmentalism, and social justice. Before addressing both pillars of our analytical approach, however, we present a short primer on the nuts and bolts of the physics of energy in the next section.

The science of energy

It is impossible to fully understand the global politics of energy from the perspective of any single scientific discipline. In reality, one needs to understand a little bit of geology, physics, economics, law, political science, and other disciplines. Getting to grips with some of the basic energy concepts and units, as well as the basic science of climate change, is essential for anyone who wants to meaningfully engage in energy studies. This section therefore offers an introduction to the key terminology commonly used to discuss energy systems. We will discuss the basics of climate science in chapter 5. We have taken care to explain the basic lexicon of energy concepts and units in accessible, non-specialist terms.

Let us begin with the term "energy" itself. When thinking of energy, most people probably think of oil, coal, natural gas, and electricity – in short, something for which they get a bill at the end of the month or when they fill up their car. From a hard science perspective, however, energy is much broader than that.[21] The word "energy" derives from the ancient Greek word *energeia*, a term coined by philosopher Aristotle, which literally means "activity in work." It describes the process of producing change (of motion, temperature, composition) in an affected system (an organism, a machine, the planet).[22] Those processes include things as different as muscle power of humans and animals, wind blowing through sails, or burning oil to warm one's home.[23]

These examples show that energy comes in many forms. Its most commonly encountered forms are heat (thermal energy), motion (kinetic or mechanical energy), light (electromagnetic energy), and the chemical energy of fuels and food. Energy can be converted from one form into another. For example, the sun sends out electromagnetic energy (light), which is converted by plants into stored chemical energy (sugar). Some of these plants are consumed by animals and converted into mechanical energy (the horses that pull a wagon). In other cases, the energy remains stored in the plant for thousands of years, eventually becoming a fossil fuel (coal, oil, or gas) that might be converted into thermal energy through combustion.

No energy is lost, or disappears, as energy is converted from one form to another. This principle comes from the first law of thermodynamics, the "law of conservation," which says that the total amount of energy in the universe remains constant. Energy cannot be created or destroyed; it can only be converted from one form to another. A physicist would thus never use the words energy "consumption" or "production," yet these terms are commonly used by laypersons and energy experts alike and in this book, too, we will stick to the conventional terminology.

Although energy is always conserved across conversions from one form to another, it becomes less useful. This is the second law of thermodynamics, the law of entropy, which says that there is a natural tendency for things to degenerate into increasing disorder. Imagine you wake up in the morning and start baking an omelet only to realize you want oatmeal instead. You cannot put the broken eggs back into their original shells. That is the principle of entropy. In closed systems, things go from low entropy (more order) to high entropy (more disorder) – never the opposite. A simple way to grasp the meaning of entropy is to say that all processes of change (or conversion) are irreversible. A lump of coal is a highly ordered (low entropy) form of energy. When combusted, it produces heat, a dispersed (high entropy) form of energy. The sequence is irreversible: the heat and gases that are released during the burning of coal cannot ever be reconstituted as a lump of coal.[24] All exchanges of energy are subject to inefficiencies, such as friction or heat losses, which increase the entropy of the system. Thus, the efficiency of any energy conversion process will always be less than 100 percent. One vital corollary of this fundamental truth is that energy, when wasted, cannot be used for its intended purpose. Thus, energy lost as waste-heat in an old-fashioned incandescent light bulb cannot be used by real people ("end use customers") for desired goals such as light, motive power, or cooling. A car that has run out of gas will not run again until you walk 10 miles to a gas station and return with petrol to refuel the car.

These two laws are incredibly important for comprehending how our global energy system works.[25] Some even go as far as to say that the laws of thermodynamics condition our economic and political systems and determine the rise and fall of civilizations.[26] As chemist and Nobel laureate Fredrick Soddy wrote in 1911, the laws of thermodynamics:

> control, in the last resort, the rise or fall of political systems, the freedom or bondage of nations, the movements of commerce and

industry, the origin of wealth and poverty, and the general physical welfare of the race.[27]

According to the international system of units, or metric system, the basic unit of energy is the joule (J), named after the English physicist James Prescott Joule, whose work led to the development of the first law of thermodynamics. One joule represents the work done by a force of one Newton traveling over a distance of one meter. Joule is not a very convenient metric for expressing the total energy use in many practical applications, such as when calculating the monthly utility bill, since it gives you very large numbers. Therefore, other metrics are more commonly used, such as the calorie, the British thermal unit (BTU), the kilowatt-hour (kWh), and tons of oil equivalent (toe). A calorie is defined as the energy required to raise the temperature of 1 gram of water 1 degree Centigrade. Data can be easily converted from one metric into the other. For example, 1 calorie is equivalent to 4.186 joule.[28]

While it is easy to convert one metric into another, this does not imply that all forms of energy are equivalent and interchangeable. As the second law of thermodynamics has taught us, energy can be converted from one form to another, but conversion entails losses. Fossil fuel-powered power plants, for example, transform *chemical energy* into *electrical energy* with an efficiency of 40 percent or so. That is why some statistics of energy production and consumption introduce multipliers when all the different forms of energy are put in the same units. Electricity from hydropower, for example, is typically rated as being worth 2.5 times more than the chemical energy in oil. In other words, 1 kWh of hydroelectricity equals 2.5 kWh of oil. The reason is that if you put that much oil in a standard thermal power station, it would deliver about 40 percent of 2.5 kWh, which is 1 kWh of electricity.[29]

The electricity industry uses its own vocabulary, which is worth discussing as it allows us to make a distinction between power and energy, two concepts that are often conflated. The metric unit of power is the watt (W), after the Scottish engineer James Watt, who helped to develop the steam engine in the late eighteenth century. One watt is equal to one joule per second. Note that the definition of power includes time: a 100-watt incandescent light bulb uses 100 joules of energy *every* second. The higher the wattage, the brighter the light, but also the more energy it uses. These distinctions of scale have an importance that is far greater than their simple differences in appearance when printed out on a page. For example, a measurement in watts would be typical for quantifying the energy use of a household appliance, such a 100 W incandescent light bulb. A measurement in kilowatts (kW) might most often be used to describe the instantaneous peak demand of an entire household. A measure of some level of megawatts (MW) would be typical for the total instantaneous demand of a neighborhood or a small town of a few thousand households (or for describing the output of a "small" power plant, sized and built to serve that scale). Discussions of gigawatts (GW) (each GW being a thousand MW) would be usual if describing the instantaneous cumulative power needs of large cites or provinces or states made up of millions of households. And, at a level a thousand-fold larger, a discussion of terawatts (TW) would be appropriate for describing the "single-moment" needs of a highly developed continent such as North America or Europe.[30]

Power is about the rate with which we use or produce energy, but to say something about the total amount of electricity we use or produce, we have to switch metrics. Electrical energy is typically expressed in a watt-hour and its multiples: kilowatt-hours (kWh), megawatt-hours (MWh), gigawatt-hours (GWh), and terawatt-hours (TWh), each one being a thousand-fold larger than the one before. Talking about a gigawatt-hour of energy use does not imply the energy was used *in one hour*. You could use a gigawatt-hour of energy by switching on one million toasters with a capacity of 1,000 W for one hour, or by switching on 1,000 toasters for 1,000 hours.[31] This distinction is often framed as one between the installed capacity of a system (measured in W) and the energy it generates over time (measured in Wh).

These two metrics allow us to describe the difference between two windmills, each with a capacity of 1 MW, but a utilization rate of 10 percent versus 30 percent. The difference in utilization rates could be due to the fact that one windmill is located in a place where there are more consistent winds than the other one. In that situation, both windmills have the same capacity (or power) but one generates three times as much energy throughout the year as the other.

To conclude this section, it is important to make a distinction between *primary energy sources* and fuels, which are available in nature, and *energy carriers*, which have to be produced. For example, crude oil can be found in nature but has to be processed in refineries to produce fuel products such as gasoline and diesel. Electricity and hydrogen are also energy carriers that have to be produced from other, primary energy sources. Note that some primary energy sources are *renewable*, which means that they cannot be exhausted, while others are *non-renewable*. Fossil fuels are classified as non-renewable energy

Table 1.1 Primary energy sources and energy carriers

Primary energy sources				Energy carriers
Non-renewable	Fossil fuels	Crude Oil	Transport and conversion	Electricity, hydrogen, gasoline, kerosene, nuclear fuels, etc.
		Natural gas		
		Hard coal		
	Mineral fuel	Uranium		
Renewable		Solar radiation and warmth		Electricity, hydrogen, biodiesel, pellets, etc.
		Wind		
		Water		
		Biomass		
		Geothermal warmth		

Source: Authors.

sources because the time required to form them through natural processes is measured in the tens of millions of years. At our current rate of use, existing stocks of fossil fuels are being depleted much quicker than they can be replenished through natural processes. In contrast, photovoltaic solar panels rely on sunlight, which is not depleting and will still be available in thousands of years regardless of how much we "consume" of it now. Table 1.1 gives a more complete overview of these primary sources and carriers.

A systems perspective on energy

The book encourages students to engage in *"systems thinking."* In its most abstract conceptualization, a systems approach is about identifying the interactions between elements of a whole system to better comprehend, and perhaps change, the system itself.[32] Without such a systemic focus, critical components and synergies could be missed, distorting our view of system properties and obscuring implications for efficiency or sustainability.[33] A systems approach must tackle *multiple* objectives as "a process through which the interconnections between subsystems and system are actively considered."[34]

With this goal in mind, we conceptualize energy as a "socio-technical system" – comprising not just the energy sources and technologies, but also user practices, cultural meanings, infrastructure, and supply networks. The energy system cuts across other systems such as electricity, transport, buildings, industry, and agriculture, to name a few. A systems approach involves identifying the characteristics of the system in question – its elements, interconnections, and overall function – and examining the interactions between them. To give readers a useful background that will help them navigate the rest of the book, we will break down this complex "system of systems" into three layers: the supply infrastructure, the demand infrastructure, and the social infrastructure.

The first layer, the *supply infrastructure*, includes the primary and secondary energy resources as well the systems of conversion and transportation. It encompasses the mining of coal and the extraction of crude oil and natural gas, as well as (occasionally) the mining and processing of uranium. The extractive industries also provide the material inputs – copper, rare earth elements, alumina, and others – needed to manufacture power plants, cars, transmission lines, and other electronic devices, something we call "critical materials." The supply infrastructure also involves the networks of power plants, oil and gas refineries and petrol stations, and other infrastructures that convert primary resources – including fossil fuels as well as alternatives – into energy carriers such as electricity, heat, mechanical energy, or liquid fuel. Delivery infrastructure such as pipelines, tankers, and electric transmission and distribution lines are also part of the supply infrastructure.

Standard accounts of energy security often stop here and restrict their focus to the supply infrastructure. The International Energy Agency (IEA), for example, defines energy security as the "uninterrupted availability of energy sources at an affordable price."[35] This brings into attention the oil fields, coal mines, solar farms, power plants, refineries, tankers, pipelines, the electric grid – in short, the hardware of the supply infrastructure. It could lead one to believe that making an energy transition is simply about switching from one energy source to another – say, closing down a coal-fired power plant and replacing it with windmills and solar panels.

The concept of "energy services" is helpful in broadening the perspective. Strictly speaking, there is no demand for energy. People are not interested in buying a barrel of crude oil, a ton of coal, or a cubic meter of natural gas. Instead, they want the services that energy can deliver such as mobility, lighting, refrigeration of food and medicines, heating of homes, cooking food, information and communication, and others – hence, the importance of end-use technologies or the so-called "prime movers" such as cars and their combustion engines, lamps, electric appliances and furnaces. This leads us to consider another layer beyond the supply infrastructure.

The *demand infrastructure* encompasses those end-use prime movers and the built environment, as well as patterns of energy consumption, use, or practice. Prime movers are the technology that converts primary and secondary fuels into useful and usable energy services. Without prime movers, all of the dazzling technological advances human civilization has made over the past millennia would remain nothing more than unrealized concepts. Human muscles are the classic prime movers; those muscles enabled us to hunt, gather, and farm. The first mechanical prime movers were simple sails, water wheels, and windmills; the Industrial Revolution had its steam engines and turbines; the modern area has internal combustion engines, jet turbines, compact florescent light bulbs, and household electric appliances. Yet, the demand infrastructure is larger and also comprises the way we raise buildings, how we design and integrate cars, how we organize freight corridors in our economies, how we build and design cities, and how we are able to communicate with our appliances through IT.

Expanding the horizon to include the demand infrastructure leads to conclusions that might be surprising for those with an interest in the geopolitics of energy. In press reports and public debates, discussions on the geopolitics of oil and gas typically revolve around things such as pipelines, "chokepoints" on tanker routes, and access to oil fields. This is short-sighted, as the following quote by Walt Patterson illustrates:

> Key competitors for ExxonMobil are not Shell nor BP but Honda and Volkswagen. Competitors for Gazprom are Europe's manufacturers and installers of thermal insulation. Competitors for EdF and E.On are the manufacturers of compact fluorescent and LED lamps.[36]

Would you have thought that the insulation of buildings is the best line of defense against Russia's alleged "gas weapon"?

Bringing in the demand infrastructure also has implications for how we define an "energy transition." It is no longer just about scrapping a coal-fired plant here, and erecting a windmill there. The whole demand infrastructure needs to be in the picture too. From a consumers' perspective, it might not seem to make much of a difference to switch from a petrol car to an electric one, especially if the latter reaches cost parity and driving ranges comparable to a petrol car. And yet, an electric vehicle that runs on a battery has a whole different supply chain than a petrol car with an internal combustion engine. Similarly, scrapping a coal-fired power plant and replacing it with windmills disrupts the traditional business model of the utility – something that the German utility companies RWE and E.ON have learned the hard way over the past couple of years. In short, an energy transition brings about changes on both the supply and the demand side.

Third, the *social infrastructure* captures the less tangible aspects of our energy systems. It can refer to our economic growth models, standards of living, tax rules and interest rates. It encompasses the laws, rules, and regulations that bear on our energy production and use. It covers our habits, norms, cultures, and values which all shape our energy systems in ways that are often overlooked. Things like mobility patterns, driver preferences, lifestyles, and routines.

Figure 1.3 schematically summarizes this three-tiered view of the energy system. It captures both the hardware of the system on the supply and demand side, as well as the software – that is the norms, cultures, and rules that pervade the energy system.

Contested frames

As a second distinctive feature, rather than confining the analysis to one disciplinary box – say, political science theories or economic models – this book builds an analytical framework that combines the tools and insights of political science, economics, development studies, environmental studies, political geography, and sociology. The framework is based on four archetypical "frames" through which energy is most often depicted: neo-mercantilism, market liberalism, environmentalism, and egalitarianism (see table 1.2). These frames are implicitly mentioned within every chapter, and we come back to them explicitly in the Conclusion.

In the broadest sense, a frame is simply a way of organizing experience.[37] Lakoff defines frames as "mental structures that shape … the way we see the world, the goals we seek, the plans we make, the way we act, and what counts as a good or bad outcome."[38] For Benford and Snow, framing is about actively constructing meaning, denoting "an active, processual phenomenon plies agency and contention at the level of reality construction."[39] Frames allow a social reality to be constructed or challenged; it is not merely language or words, but a reality for how actors and objects are made meaningful, and then

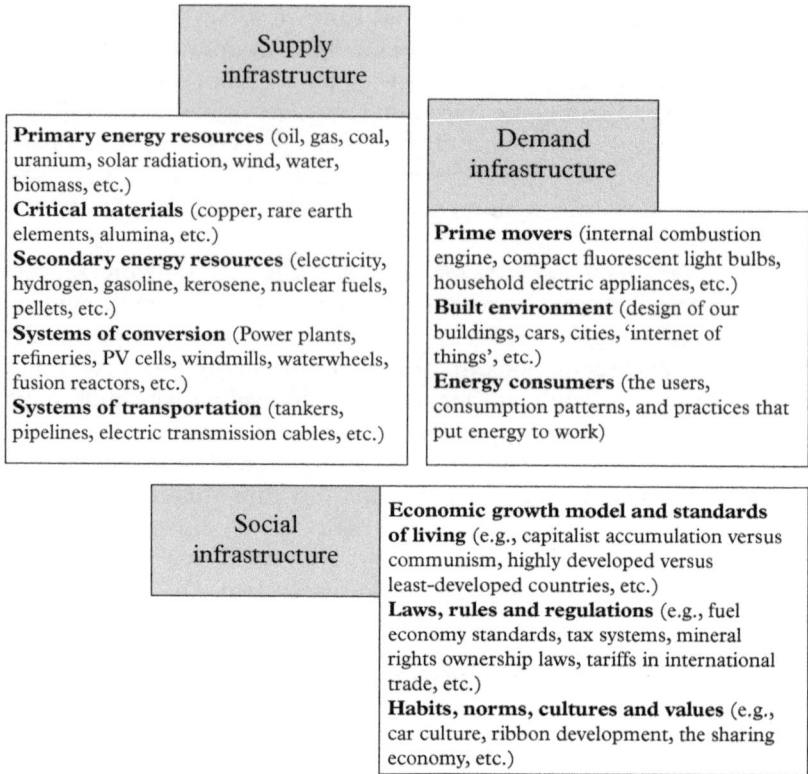

Supply infrastructure	
Primary energy resources (oil, gas, coal, uranium, solar radiation, wind, water, biomass, etc.)	

	Demand infrastructure

| **Primary energy resources** (oil, gas, coal, uranium, solar radiation, wind, water, biomass, etc.)
Critical materials (copper, rare earth elements, alumina, etc.)
Secondary energy resources (electricity, hydrogen, gasoline, kerosene, nuclear fuels, pellets, etc.)
Systems of conversion (Power plants, refineries, PV cells, windmills, waterwheels, fusion reactors, etc.)
Systems of transportation (tankers, pipelines, electric transmission cables, etc.) | **Prime movers** (internal combustion engine, compact fluorescent light bulbs, household electric appliances, etc.)
Built environment (design of our buildings, cars, cities, 'internet of things', etc.)
Energy consumers (the users, consumption patterns, and practices that put energy to work) |

Social infrastructure	**Economic growth model and standards of living** (e.g., capitalist accumulation versus communism, highly developed versus least-developed countries, etc.) **Laws, rules and regulations** (e.g., fuel economy standards, tax systems, mineral rights ownership laws, tariffs in international trade, etc.) **Habits, norms, cultures and values** (e.g., car culture, ribbon development, the sharing economy, etc.)

Figure 1.3 Anatomy of the global energy system

Source: Authors.

naturalized.[40] Furthermore, frames not only compete for public attention or "consumption"; they also compete in "framing contests" where "different social actors construct rival understandings of contested social phenomena and seek to mobilize support for their preferred 'frame' over rival 'counter-frames'."[41]

Our own analytical framework is essentially constructivist and serves to capture the complexity and diversity of individual views on global energy issues. Each frame has a different take on what is happening, what is causing it, and what can be done in the world of energy. They incorporate foundational assumptions about how the world of energy works. These frames largely structure the first part of the book.

The first frame, *neo-mercantilism*, assumes that states are key actors, aiming to maximize their power and autonomy. Energy is considered

Table 1.2 Frames and worldviews in global energy politics

Frames	Dominant worldviews	Prioritized component of energy security	Energy security for whom?	Underlying values and goals
Neo-mercantilism	Defense of national security	Geopolitical availability	State	Political independence and territorial integrity
Market liberalism	Technological optimism, free market libertarianism	Economic affordability	Economy	Welfare, freedom
Environmentalism	Environmental preservationists, conscientious consumption	Environmental sustainability	Earth	Respect for nature
Egalitarianism	Justice, neo-Marxism, feminism, equality	Social acceptability	Society	Equity, justice

Source: Van de Graaf, T. & Zelli, F. (2016). Actors, institutions and frames in global energy politics. In *The Palgrave handbook of the international political economy of energy* (pp. 47–71). Palgrave Macmillan, London, reproduced with permission of Palgrave Macmillan.

a strategic good, vital for national security and prosperity. It can be a source of power for those that have it or a source of vulnerability for those that lack it. The national interest of states revolves around securing supplies of energy, particularly fossil fuels, whose reserves are geographically concentrated. States and strategic planners worry mostly about availability of and access to such scarce reserves, which results in a geopolitical power game between net energy exporters and importers. States often deploy strategies of energy statecraft, that is, using energy as a tool in foreign policy. Sometimes this geopolitical bickering results in conflict. This frame is akin to the realist school of thought in international relations.

The second frame, *market liberalism,* looks at other actors beyond the state, including firms, non-state actors and international organizations. It essentially sees energy as a commodity just like any other. Cross-border energy trade creates interdependencies, which lower the risk of conflict, and brings benefits to all. Attempts to weaponize energy trade are rare and ultimately self-defeating, as most international energy transactions take place within a framework of well-established markets and institutions. Market regulation trumps energy statecraft. This frame stresses the dimension of "affordability," which means not only low prices for energy consumers but also stable prices to increase planning and investment security for energy producers. What is also often espoused within this frame is the belief that technological innovation progresses and continually improves the lives of people.

The third frame, *environmentalism*, puts the value of environmental sustainability front and center, referring to both the protection of the natural environment and preventing the full depletion of non-renewable energy sources by making a timely switch to renewable and other low-carbon energy sources. Mainstream environmentalism believes that environmental protection does not require fundamental changes in the political and economic structures of modern society, whereas a more radical variant sees the capitalist model of economic growth as a major cause of environmental stress. This can involve not only climate change and greenhouse gas emissions, but changes in land use, impacts on water availability and quality, radioactivity, and the spread of toxic flows of pollution.

The fourth frame, *egalitarianism*, puts the spotlight on equity, equality, and justice. In the Marxist variant, energy resources are exploited for the advancement of capitalist classes. Western states and multinational enterprises have close bonds with local elites in resource-rich countries and the wars in the Middle East reflect the need of capitalist states to access petroleum. The key global energy challenge in this perspective is the massive inequality embedded in our energy system, with the average American consuming much more than the average Nigerian, and many communities and poor countries effectively deprived of access to even the most basic energy services. This frame stresses respect for human rights and dignity in relation to both individuals and social groups, and it looks at the costs for workers and disenfranchised communities of both the current energy system, and any future energy transition.

Plan of the book

Having laid out the scope and approach of the book, the next chapter traces the political history of energy and explains the functioning of the global energy system.

The remainder of the book then proceeds in two parts. The *first part* discusses the implications of our fossil-fuel based energy system on four key areas of international relations: security, the economy, the environment, and global justice. We show that energy is a prism through which these broader issues refract. It is entangled with other major topics such as climate change, development, and foreign policy.

The *second part* examines ongoing and future changes in the global energy system. It examines technological options to move away from fossil and nuclear fuels, and the socio-technical barriers to their widespread adoption. In addition, this part looks at how energy is governed nationally (in key consumer and producer countries) and internationally (in multilateral fora).

At the end of the book, in our Conclusion, we put forward four conjectures about the future of energy. These suppose that the future politics of energy will be interdependent, interconnected, contested, and uncertain. No country has the privilege of becoming disentangled from global energy interdependencies for fuel, technology, knowledge, or even the consequences of actions or crises in other countries. Energy remains a complex issue that is intertwined with other important social and political issues including employment, environmental sustainability, fiscal policy, and public health. Because energy issues involve producers vs. consumers, importers vs. exporters, rich countries vs. poor countries, and local vs. national and even transnational interests, it will remain a key site of contestation and disagreement. Finally, the transformation currently ongoing in the energy sector has such a degree of complexity and rapid change that it is inherently unpredictable. Herein lies perhaps the most intriguing aspect of global energy politics: despite so much being at stake, the future is truly dependent on the choices we will collectively make within the next few decades. The future is myopic, but also mutable and thus potentially munificent.

2

The history and functioning of energy markets

This chapter introduces the reader to the history and functioning of key energy markets, especially oil and gas, which are the most internationalized segments of the world energy business. By tracing the history of each major energy source and industry, we can get a better sense of how the current market structures and pricing regimes came into being, and what alternatives have preceded it. Our focus lies on the international political economy aspects of these energy sources – that is, the trade patterns, international value chains, general market and industry structure, and the political risks and dependencies that have developed around these energy sources.

The chapter starts with a section on oil, the first commercial fuel to be traded internationally in large quantities, and one with a complex history that may be unfamiliar to many students or political scientists. Today, oil is still by far the single largest commodity in international trade and it is one of the largest and most international of all industries in the world. Next, we move on to discuss natural gas, a fuel that is proportionally less traded internationally but nevertheless generates particular types of dependencies. We also discuss the international markets for coal, biomass and uranium, before discussing the international dimensions of various types of renewables.

Oil and the internationalization of the energy sector

Oil is, for at least the foreseeable and immediate future, vital to the functioning of the world economy, accounting for one-third of global energy consumption. While petroleum is used across a

Table 2.1 Major oil reserve holders, producers and exporters (2017)

	Share in reserves		Production (mb/d)		Exports (mb/d)
Venezuela	17.9%	US	12.0	Saudi Arabia	8.3
Saudi Arabia	15.7%	Russia	11.2	Russia	5.1
Canada	10.0%	Saudi Arabia	11.1	Iraq	3.8
Iran	9.3%	Iraq	4.5	US	3.2
Iraq	8.8%	Iran	4.0	Canada	2.7
Russia	6.3%	China	4.0	UAE	2.7
Kuwait	6.0%	Canada	3.7	Nigeria	2.3
UAE	5.8%	UAE	3.1	Angola	1.7
US	6.0%	Kuwait	2.9	Kuwait	1.7
Libya	2.9%	Brazil	2.5	Venezuela	1.5

Notes: All data are for 2017. Production and exports data include crude oil, oil products, natural gas liquids and biofuels. Reserves are proved reserves – generally taken to be those reserves that can be recovered with reasonable certainty under current economic conditions.

Source: Reserve data from BP (2019) *Statistical Review of World Energy*; data on production and exports come from EIA, https://www.eia.gov/beta/international/.

range of sectors, it is absolutely indispensable for the movement of people and goods, supplying no less than 92 percent of energy for transportation.[1] Without oil, our modern civilization would literally come to a standstill. All countries in the world are oil consumers, but oil reserves are quite concentrated (see table 2.1). Even though we talk about the global oil price, crude oil is actually not a uniform commodity. Depending on its chemical composition, crude oil can be either "sweet" or "sour" depending on its sulfur content; or "heavy" or "light" depending on its gravity. These distinctions matter because many refineries can only process one type of crude oil.

Still, economists like Nobel prize winner William Nordhaus have proposed a "bathtub model" of the oil market.[2] Since oil is largely fungible, you can imagine the oil market as a giant bath tub with taps filling it from all the producers, and drains emptying it from all the consumers. In this simple but powerful image, it does not matter where oil comes from, international oil prices tend to converge, and embargoes on imports or exports of oil are impractical. Yet, the oil market did not always function as a "bath tub" or global auction. This section explains how the industry has developed over the past 150 years into the giant world business it is today.

The birth of an industry

Petroleum has been known, and used, since antiquity, but the modern oil industry is commonly said to have started in 1859. In that year, Edwin Drake drilled the first oil well in a place called Titusville, Pennsylvania. This marked the start of large-scale commercial exploitation of oil. In this period, the main commercial use of petroleum was for lighting, as kerosene came to supplant whale oil as a fuel for lamps and cooking. A popular argument that is often made is that the oil industry saved the whale from extinction although, clearly, the growing use of oil did not end whaling.[3]

Drake's drilling of the first oil well triggered what could be called a "black gold rush." The fields of West Pennsylvania suddenly attracted thousands of amateur petroleum seekers, nicknamed "wildcatters."[4] Two unique laws in the US helped to spur the development of the oil industry. The first pertained to mineral rights ownership. In most countries in the world, the government retains the ownership over the minerals, including oil and gas, on its territory. In the US, if you own a parcel of land, you are also entitled to the sub-surface minerals that lie beneath it. So, your ownership rights extend vertically downward from your property. The second law has come to be known as the "law of capture." The rule of capture provides that anyone who drills oil from his land is allowed to drain out that oil, even if some of that oil originated from beneath a neighbor's land. In a sense, it is a horizontal extension of the mineral ownership rights beneath the surface. The law of capture gave landowners a strong incentive to pump out oil as quickly as possible, leading to considerable waste.[5]

It was in this early period of chaotic competition that John D. Rockefeller would succeed in consolidating the industry and establishing a monopoly. His strategy was to first take over all of the so-called downstream structures of the oil market, such as refineries and all of the common transportation routes for oil (pipelines, railroads, ships, and so on). By the early 1880s, his company, Standard Oil, controlled about 90 percent of US refineries and pipelines. Only at a later stage did Standard Oil get into the other critical part of the business: the oil production itself. There, too, it would soon become a dominant force. This whole system was broken down in 1911, when the US Supreme Court ordered the dismantling of Standard Oil for violation of anti-trust laws. Rockefeller's business imperium was then broken down into no less than 34 different companies, among them those that later became ExxonMobil, Chevron, and BP (see figure 2.1).

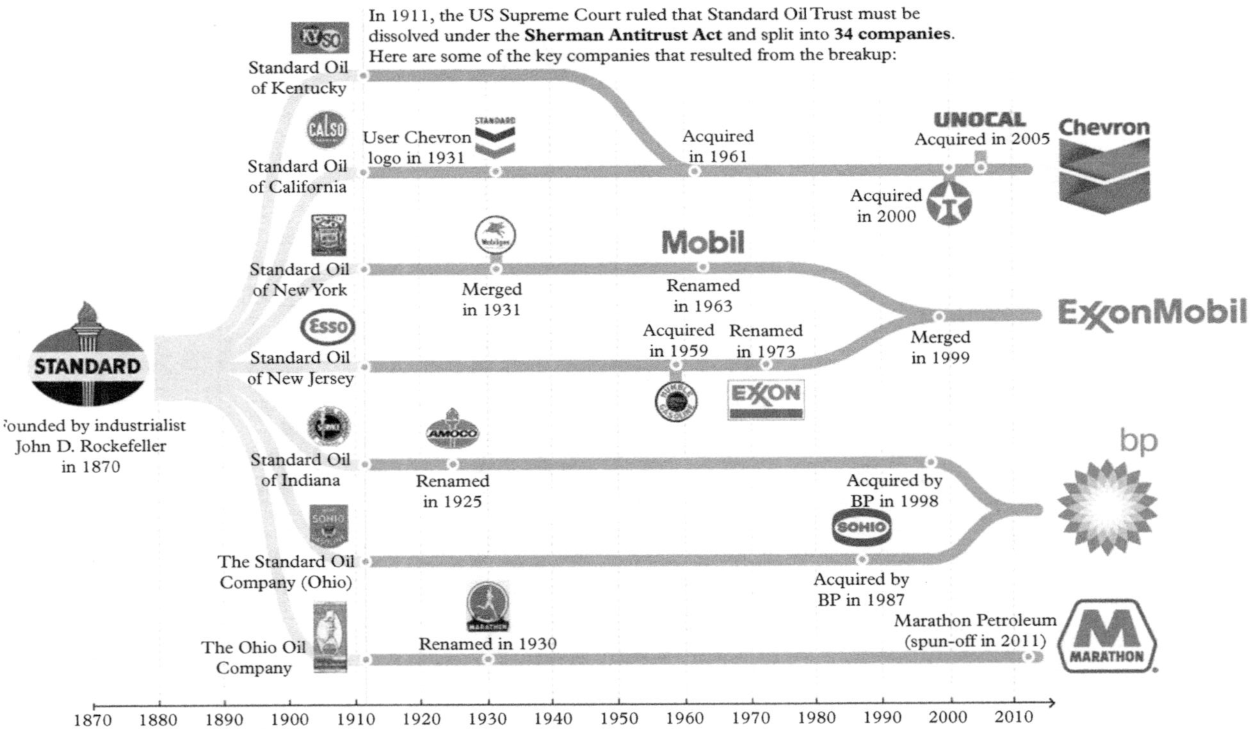

In 1911, the US Supreme Court ruled that Standard Oil Trust must be dissolved under the **Sherman Antitrust Act** and split into **34 companies**. Here are some of the key companies that resulted from the breakup:

Figure 2.1 The break-up of Standard Oil and its legacy

Source: Visual Capitalist, reproduced with permission from https://www.visualcapitalist.com/chart-evolution-standard-oil/.

Around the turn of the millennium, consumption of petroleum products for lighting and cooking declined in the face of competition from gas and electricity. Kerosene lamps had been gradually pushed out of the market by the light bulb, which Thomas Edison had patented in 1880. However, the decline of the industry was averted largely by the introduction and spread of the internal combustion engine.[6] Two key developments would elevate the economic and strategic importance of oil. First, in 1908, Henry Ford introduced the Model T, the world's first mass-produced car. The cheaply-priced Model T would change the automobile from a luxury product to one that workers could afford. This accelerated the expansion of petroleum into other end-use sectors beyond illumination, and spurred a huge increase in oil consumption.

Second, in 1912, admiral Winston Churchill decided to convert the Royal Navy from (British) coal to (foreign) oil. This was a risky decision, as it made the UK military dependent on energy imports. It led the UK government already in 1914 to acquire a majority stake in the struggling Anglo-Persian Oil Company, known today as BP. Yet, Churchill's bet gave Britain a strategic advantage during the war, when its oil-powered fleet outperformed the coal-fired German Navy in terms of range and speed.[7] World War I thus critically underscored the military importance of oil to fuel the machines of war, initially just ships, but soon also tanks and airplanes. Oil, for the first time, became a strategic commodity whose continuous supply would be essential for the security of nations. After the war, all major economies became drawn into a quest for oil.[8]

Era of the Seven Sisters

The postwar struggle for oil led to the creation of an imperial system of concessions that would dominate the international oil trade until the 1970s.[9] Much of the struggle focused on the Middle East where the major powers, especially Britain, France, and the US, and their international oil companies fought over access to oil reserves, or denial of access to others. The League of Nations had given Britain and France very significant mandate powers over Mesopotamia, a region centered around present-day Iraq that was thought to be highly prospective of oil.[10] The British and French tried ruthlessly to keep the American firms out of the Middle East which, in the words of Adam Sampson, "they regarded as their own equivalent of Texas."[11]

After several years of confrontation and competition, in 1928 two key agreements were finally reached. Under the so-called "Red

Line Agreement," the major companies effectively divided Middle East oil among themselves. More precisely, the five companies that constituted the Iraq Petroleum Company agreed not to engage in competitive bidding wars in a region roughly coinciding with the former Ottoman empire. A few months later, the major companies signed another deal at Achnacarry Castle in Scotland in which they divided up shares in the global market to avoid destructive competition between them. This agreement became known as the "As Is Agreement," as the companies agreed to keep their shares of oil exports "as is" and not compete for more.[12] It also linked international oil prices to US domestic prices, a system that both protected the more expensive American oil and was very lucrative for the international majors.[13] The Achnacarry agreement remained secret until 1952.

Through these agreements, eight firms came to dominate the bulk of international petroleum trade over the next decades. Five were American: Exxon, Mobil, Texaco, Gulf, and Chevron. Two were Anglo-Dutch: BP (entirely British-owned) and Royal Dutch Shell (jointly owned by British and Dutch interests). One was French, the Compagnie Française des Pétroles (CFP), the ancestor of Total, which was much smaller than the others. The seven US and Anglo-Dutch firms eventually came to be called the "Seven Sisters."[14]

Each of them soon developed into "vertically integrated" oil companies, controlling virtually all non-communist oil from well to pump. They were active along the entire value chain, from the "upstream" business of drilling and producing at oil fields, to the "downstream" activity of distributing and selling at the pumps. And each company tried to be self-sufficient at both ends, so that "their oil could flow into their tankers through their refineries to their filling stations."[15] This led to a situation in world oil trade where most of the crude that moved internationally was not sold at all. It was simply transferred between subsidiaries of the same parent company. It did not change ownership until it was sold at the pump or factory – the point at which all oil, finally, changed owners.[16] Since there was no free market operating outside of these inter-company exchanges, there was no international oil price. The oil companies used a "posted price" to calculate the royalties and tax income accruing to host governments.[17]

With these two agreements in place, the oil majors had effectively cartelized the international oil market. The whole system was threatened, however, by the giant oil discoveries in East Texas in 1931 which, together with the Great Depression, lowered oil prices in the US. In response, a state body, the Texas Railroad Commission, would begin to regulate and control output in this key oil-producing state.[18] Restricting supply in Texas, the key oil frontier in the US, was also necessary to maintain global oil prices, which were linked to US prices in the As Is Agreement. Some three decades later, the founding fathers of OPEC would look back at the Texas Railroad Commission as a model for their cooperation.[19]

The OPEC revolution

After World War II, the Seven Sisters cartel began to break down in parallel to the collapse of colonial power. From the late 1940s onwards, the oil-exporting countries began to negotiate increasingly favorable deals. In an attempt to capture a larger share of the oil rents, they forced the oil majors into renegotiations of royalties, taxation, and cost-sharing schemes. The first challenge to this concessionary system came from Venezuela, which adopted a 50/50 profit-sharing law in 1948. The idea spread rapidly to the Middle East and, by 1952, all the major countries in the region had switched to a system of profit taxes.[20] Iran went one step further and nationalized its oil industry, seizing BP's assets, in 1951. After two years of sanctions against Iranian oil, and amid US fears that Iran would fall prey to communism, the CIA orchestrated a coup that led to the overthrow of Iranian Prime Minister Mossadegh in 1953.[21]

But nationalist sentiment in the region could not easily be contained and in 1956, Egyptian President Gamel Abdel Nasser nationalized the Suez Canal, seizing another British asset that symbolized an imperial past. The crisis brought commercial oil flows through the canal to a halt, leading to severe oil shortages in Western Europe, a situation the press dubbed "Europe's oil famine."[22] It also triggered a military response from the UK, France, and Israel. US President Dwight Eisenhower, wary that the conflict would play into Soviet hands, declined to send additional oil shipments to needy European countries until the British and French had withdrawn their troops from the area. The oil-starved British and French quickly succumbed to American pressure and, a year later, the canal was reopened.[23]

Between 1945 and 1970, oil demand grew rapidly on the back of the "economic miracle" in the Organisation for Economic Co-operation and Development (OECD) countries and, in 1965, oil surpassed coal as the world's chief energy source. Oil production in the non-communist world increased from 8.7 million barrels per day (mb/d) in 1948 to 42 mb/d in 1972.[24] Output increased, especially in the Middle East and North Africa, which became leading centers for the production and export of oil. After World War II, the US was still the world's largest oil-producing country but its consumption also rose rapidly and, by 1948, it became a net importer. To shield domestic oil producers in the US from the flood of low-priced foreign oil, the Eisenhower administration put in place oil import controls in 1959. As a result, the international oil companies cut the posted prices in 1959 and 1960, resulting in lower income for host governments.[25]

In an attempt to prevent declines in the "posted price," the oil-exporting countries began to organize themselves. In September 1960, the Venezuelan Oil Minister Pérez Alfonso and the Saudi Oil Minister Abdullah Tariki took the initiative to create the Organization of the Petroleum-Exporting Countries (OPEC). Apart from Venezuela and Saudi Arabia, other founding members were Iran, Iraq, and Kuwait. By uniting, these countries hoped to stand stronger in the negotiations with the international oil companies on royalties and tax questions.[26] In its first decade, OPEC functioned more as a "trade union" than as a cartel.[27]

By the late 1960s, the oil market significantly tightened on the back of rapidly growing oil demand. Due to geological depletion, the United States' spare capacity had effectively disappeared by 1971. As a result, the United States could no longer operate as a swing producer to offset oil supply disruptions. By 1973, OPEC had 12 members and was producing 53.9 percent of total world oil output. Against this background, the oil-exporting countries would succeed in slowly wresting control of production and prices from the oil majors.

The lead was taken by Libya where Colonel Gaddafi, after seizing power in 1969, threatened to expropriate any foreign oil company that did not cut production and pay more taxes. The oil companies reluctantly gave in, and the Libyan example was followed by other oil-producing states, all claiming a larger share of the companies' profits. After a while, the companies united in a common front and sought to negotiate with OPEC as a bloc. This resulted in the 1971 Tehran and Tripoli Agreements, which increased royalties and posted prices. Simultaneously, there was a wave of nationalizations in the oil industry, including in Libya (1970), Iraq (1972), and Venezuela (1974), which led to the present era in which national oil companies control the majority of the world's oil reserves.

OPEC's dramatic assertion of market power in the early 1970s would fundamentally alter the rules of the international oil game.

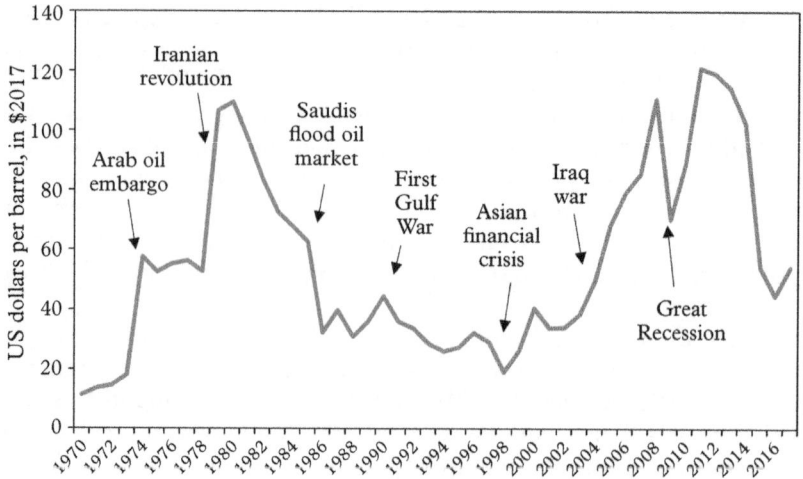

Figure 2.2 Global oil prices and major political events, 1970–2017 (in $2017)

Source: BP (2019) *Statistical Review of World Energy.*

After taking control over their national oil and gas sectors, the OPEC countries were now able to set world oil prices throughout the 1970s, using Arabian Light as a benchmark and defining reference prices for member states' oil exports.[28] The nationalizations also broke up the vertical integration of the oil industry. International oil companies were now deprived of their direct control over oil reserves, and had to engage in trading with the newly formed national oil companies to secure access to oil.[29]

These developments set the stage for OPEC to discover the political utility of its "oil weapon." The trigger to the oil crisis was the Yom Kippur War of October 1973. The pro-Israeli position of the United States and the Netherlands in that conflict prompted some Arab oil-exporters – not OPEC, as is commonly believed – to impose oil embargoes on those two countries. The export ban was later extended to Portugal, South Africa, and Rhodesia (renamed Zimbabwe in 1980). As a result, oil supplies fell about 9 percent on a global scale between October and December 1973 and oil prices quadrupled.[30] Figure 2.2 shows the evolution of average annual global oil prices (adjusted for inflation) since 1970.

In response to these events, American Secretary of State Henry Kissinger convened an energy conference in Washington in February 1974. The existing institutional arrangements for addressing energy issues, mainly through the committee structure of the OECD, were perceived as incapable of decisive action.[31] Kissinger therefore wanted to create a new organization. Initially he intended to set up an anti-OPEC consumers' cartel, but the European states and Japan, which were much more vulnerable to oil supply interruptions than the United States, successfully resisted this call.[32] By November 1974 agreement was reached on the International Energy Program, establishing the IEA as an autonomous agency of the OECD.

The first major test for the IEA came in 1979. After months of demonstrations, the pro-Western Shah of Iran fled the country, never to return. Ayatollah Khomeini returned from exile and helped transform the country from a monarchy into an Islamic Republic. Iranian oil production had dwindled from 5.5 million barrels per day to 40,000 barrels per day in January 1979, leading oil prices to jump. Prices had not yet fully recovered when war broke out between Iran and Iraq in September 1980, pushing oil prices even higher. It would be their last jump, however.

The "commodification" of oil

After the two oil shocks in the 1970s came the "counter-shock" of the 1980s.[33] Oil demand collapsed (a result of economic stagnation, fuel switching, and efficiency measures) just as new supplies of oil hit the market (for example, from the North Sea and the Gulf of Mexico), sending oil prices downwards. In March 1982, for the first time, OPEC tried to introduce production quotas, much like the Texas Railroad Commission had done until the 1960s. The effort was not very successful due to extensive cheating by some OPEC members. Saudi Arabia tried to compensate for this by reducing its own production, which declined from 10.3 mb/d in 1980 to 2.8 mb/d in 1985.[34] By that time, Saudi Arabia's patience was exhausted and it decided to punish the cheaters by restoring its oil production. As a result, oil prices collapsed in 1986 and would not recover until after the new millennium. This dramatic drop in oil prices is widely thought to have contributed to the downfall of the Soviet Union and the end of the Cold War.[35]

This was also the period in which neoliberal ideas spread globally, and the oil market did not escape from these broader ideological shifts. The United States was a key advocate for these liberal policies.[36] An important step down the road of laissez-faire liberalism was President Carter's decision, in April 1979, to decontrol domestic oil prices.[37] When the Reagan administration took office, in

1981, the United States turned to a strategy of outright "laissez-faire liberalism." That is, government intervention was to be eradicated from the domestic and international oil markets altogether.[38] Following domestic deregulation, successive United States administrations have sought to steer more of the world's oil trade into spot-market pricing systems, rather than the long-term, bilateral supply contracts favored by OPEC.[39] A marginal spot oil market had existed for years, centered mainly in Rotterdam, but by 1978, it accounted for only a meager 3–4 percent of the total oil trade. Its scope was broadening, though, thanks to the behavior of many producer countries. Eager to exploit the bullish oil market, they began selling oil on the spot markets, which were then more profitable than the traditional long-term contracts that had been the rule for decades. Maugeri duly concludes that, "by a fateful irony, it had been OPEC itself that let loose the 'monster' – the market – during the 1970s, when it fell victim to its own greed."[40]

More than anything else, the launching of the first oil futures contracts – often referred to as "paper oil" – on the New York Mercantile Stock Exchange (NYMEX) in 1983 and on London's International Petroleum Exchange (IPE) in 1988 helped to craft a new oil world no longer depending on bilateral long-term contracts. As the annual volume of contracts increased year after year, these exchanges began to handle an ever-larger portion of the world's oil trade, and they became the principal arenas for price formation on the international oil markets.[41] According to Maugeri, the introduction of futures contracts "was a historical turnaround for a sector that for many decades had seen the price of oil governed by more or less successful oligopolies."[42]

Oil prices would stay low throughout the 1990s. The First Gulf War only caused a short-lived spike in oil prices. The emergence and consolidation of a global oil market marked by relatively low and stable oil prices gave way to the view that oil is a commodity just like any other commodity. This "commodification" of oil coincided with progressing globalization, and the opening up of the former Soviet Union oil and gas sector attracted a lot of foreign investors.[43]

Towards the end of the 1990s, in the wake of the Asian financial crisis, oil prices even traded at less than 10 dollars per barrel. Ironically, at their 1997 meeting in Jakarta, OPEC oil ministers had agreed to increase production by 2 mb/d to meet what they expected to be buoyant demand from Asia. Ever since, the "Jakarta syndrome" would haunt OPEC negotiators: the danger of increasing production when demand was weakening or even just uncertain.[44]

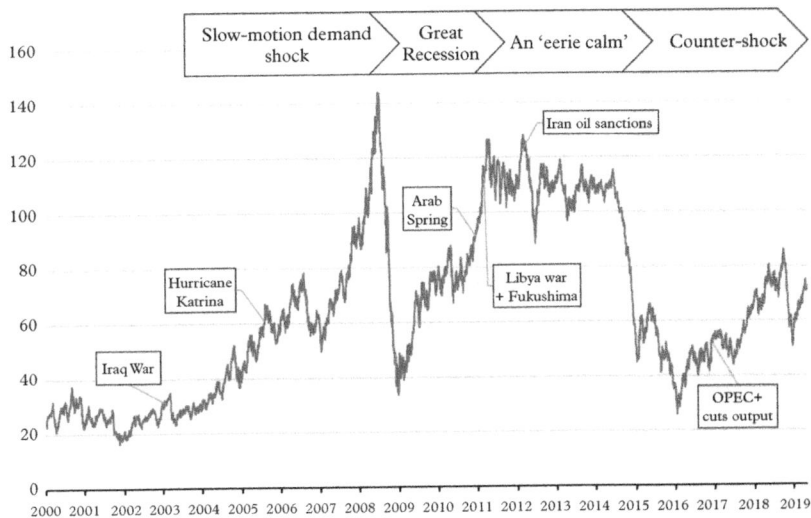

Figure 2.3   Oil price volatility and major events, 2000–2019
Note: Daily Brent spot price, dollars per barrel.
Source: EIA: https://www.eia.gov/dnav/pet/hist/RBRTED.htm.

The 1998 price collapse hit producers hard. It set off "the most far-reaching reshaping of the petroleum industry since the breakup of the Standard Oil Trust by the US Supreme Court in 1911."[45] There were several mergers between oil companies creating so-called "supermajors": BP and Amoco (1998), Exxon and Mobil (1999), Total and Elf (1999), Chevron and Texaco (2000), and Conoco and Phillips (2001). Oil-exporting countries also took a hard hit. The economic crises in Venezuela and Russia allowed new strongmen to seize power – Hugo Chavez and Vladimir Putin. In March 1999, four non-OPEC countries (Mexico, Norway, Russia, and Oman) were explicitly included in the OPEC scheme of production cuts.[46]

The era of volatility

Since the turn of the millennium, and particularly since 2004, oil markets have been extremely volatile.[47] After a long cycle of low prices in the period 1985–2003, the period 2003–14 witnessed a cycle of high (and volatile) energy prices (see figure 2.3). This shows how oil prices are doubly volatile – reacting not only to political events (as described above), but to economic fluctuations and pressures. To understand the rollercoaster oil prices of the past decade-and-a-half, it is useful to distinguish four phases.

In the first phase, roughly from 2003 to 2008, the world arguably experienced a third oil shock. Contrary to the oil shocks of the 1970s, there was no clear single disruption on the supply side. True, there were production losses due to a strike in Venezuela's oil sector, ethnic conflict in Nigeria, continued instability in Iraq after the 2003 invasion, and Hurricane Katrina in the US, which added up to a sort of "aggregate disruption."[48] Two elements, however, which were not present in the 1970s, should be added to the explanation. First, there was an unexpected "demand shock" from emerging markets, particularly China and India. As a result, world oil demand grew by almost 5 mb/d between 2003 and 2006.[49] Second, the financialization of oil amplified the imbalances between supply and demand. Trade in oil futures and options rose significantly in this period, as droves of investors and speculators came to see crude oil as an interesting investment opportunity.[50]

The result of all this was that oil prices kept on climbing until the summer of 2008, stoking concerns over energy security and "peak oil." Oil prices even set an historic record in July 2008, when a barrel of oil was trading at 147 dollars. This is a level that, even after correcting for inflation, was never attained before, not even in the turbulent 1970s. Yet, then came the interruption of the Great Recession. What started as a housing bubble in the US in late 2007 soon developed into a global financial crisis. The surge in oil prices interacted with the housing crisis to tip the US economy into recession.[51] As a result, oil prices fell off a cliff. They tumbled from almost $150 a barrel in July to less than $40 a barrel in December 2008. This second phase only proved to be a short intermezzo, however.

Oil prices recovered quite soon after the Great Recession. The price rebound was linked to OPEC's decision to cut production by 4.2 mb/d as from January 2009.[52] On the demand side, China's economy recovered quickly from the crisis, registering double-digit growth numbers in 2010, even as European and North American economies were still reeling. Turbulence in the Middle East and North Africa would further push prices higher. In December 2010, Mohammed Bouazizi, a young fruit vendor in Tunisia, set himself on fire, igniting a wave of protests in the Arab world that would eventually lead to the toppling of long-standing rulers such as Ben Ali (Tunisia), Gaddafi (Libya), and Mubarak (Egypt). Oil production in Libya was crippled, removing around 1.5 percent of supplies from the market by March 2011. As unrest continued in the Middle East, anxiety rose about further disruptions to supply, taking oil prices once again higher. This coincided with a major international crisis over Iran's nuclear program, after the discovery of secret enrichment facilities as well as efforts to develop nuclear explosive technology in the period 2010–11. It led the US and the EU to impose oil embargoes against Iran.[53]

In hindsight, this third phase could be described as a period of "eerie calm" in the oil market.[54] The apparent calm and price stability in the oil market was the accidental outcome of two opposing forces: the relentless growth in oil production in the US shale industry which, by sheer coincidence, was about as large as the huge supply disruptions in the Middle East and North Africa. Both forces neutralized each other, until in mid-2014, US oil production growth became too apparent to ignore, and oil prices spiraled down. This marked the start of the fourth phase, the counter-shock.

Just as the oil shocks of the 1970s had ushered in the development of new oil provinces, notably the North Sea and the Gulf of Mexico, so the price rises after 2000 led to an increase in exploration spending. The increased spending on exploration shifted the frontiers of drilling, notably to the deeper offshore, the Arctic, Canadian tar sands, and – most notably – US shale oil production. The US shale or fracking revolution reversed a decade-long decline in US oil production. Between 2008 and 2014, the US added no less than 5 mb/d to global production, propelling the US to become the world's biggest oil producer in 2018. The unexpected shale boom was part of a broader surge in oil and gas production. More recently, some of the established Middle Eastern producers, most notably Iraq and Iran, have increased their production after decades of wars and sanctions (see figure 2.4), adding further to the oil glut.

OPEC's response to the oil price plunge of 2014 reveals much about the limits for the club to manage oil markets. OPEC could not agree to production cuts in 2014 like it had done in 2009, in the wake of the Great Recession. Saudi Arabia still remembered its awful experience from the 1980s when it took the lead in organizing production cuts only to see oil prices falling further. This time, Saudi Arabia wanted to shift the burden of balancing the market to the high-cost producers, notably the frackers in North America. When US shale proved much more resilient than anticipated, OPEC switched strategies. In late 2016, OPEC countries and a group of 11 non-member producers (together known as "OPEC+" or the Vienna group) agreed to cut production.[55] Russia and Saudi Arabia are spearheading this cooperation, leading many observers to declare the death of OPEC.[56]

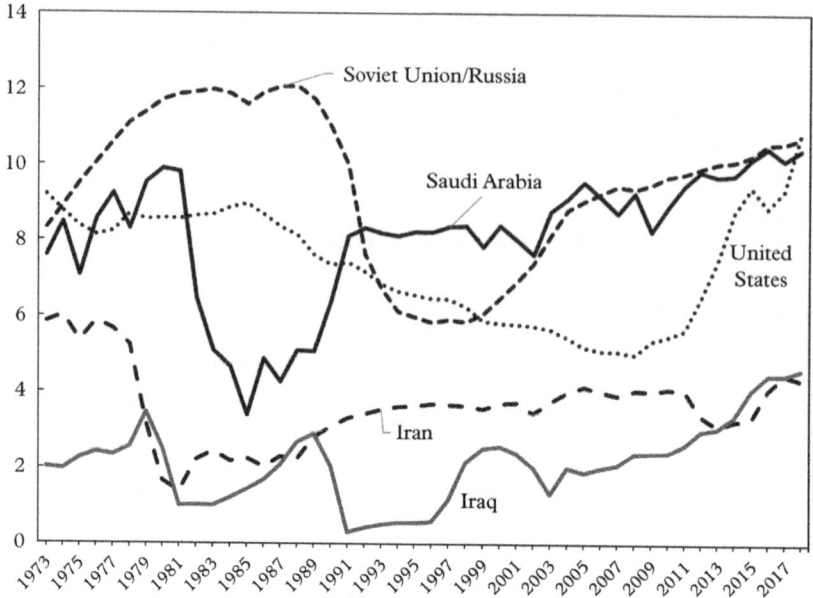

Figure 2.4   Yearly crude oil production of selected countries
(mb/d, 1973–2018)

Source: EIA (2019) *Monthly Energy Review*, April.

The incomplete globalization of gas markets

The history of the gas sector is fundamentally different from oil. The difference stems in no small part from the different chemical composition of natural gas. Natural gas, or simply gas, typically consists of 70–90 percent methane (CH_4) and 0–20 percent butane, propane, and ethane. Just like oil, it is a mixture of hydrocarbons, formed by the process of pressurizing and heating organic matter decayed under the earth's surface for millions of years. In fact, gas is often found alongside oil – called "associated" or "wet" gas – but it can also be found in fields isolated from oil reserves – called "dry" gas. The key differences with oil, however, is that gas exists in gaseous condition whereas oil is a liquid. This makes it more difficult – or, rather, expensive – to store and transport natural gas over long distances. Gas trade depends on expensive pipelines or liquefied natural gas (LNG) facilities (not to be confused with CNG or LPG, see box 2.1). As a result, natural gas has typically been traded on regional markets.

Box 2.1 What is the difference between LNG, LPG and CNG?

LPG stands for liquefied petroleum gases. It consists of a mixture of hydrocarbons and comprises propane (C_3H_8) and butane (C_4H_{10}) or a mixture of the two. These gases can easily be liquefied at room temperature by increasing the pressure. The liquid in bottles of camping gas and cigarette lighters is compressed propane and butane. LPG is also used as an alternative fuel to gasoline in cars. Because LPG vapors are heavier than air, LPG-fueled vehicles are not allowed to enter certain underground parking facilities.

LNG stands for liquefied natural gas. It is natural gas (CH_4) that is turned into a liquid by cooling it to $-162°C$. This results in a 1/600 volume reduction, and allows the gas to be stored on LNG tankers or used as a fuel for heavy-duty trucks or even ships.

CNG stands for compressed natural gas. It is natural gas (CH_4) that is put under high pressure, resulting in a 1/100 volume reduction. CNG can be used as a fuel for internal combustion engines that have been modified for CNG use or in vehicles that were specifically manufactured for CNG use.

From town gas to natural gas

The first commercial use of gas was so-called "town gas," which was natural gas manufactured from coke and coal, often located close to communities within local reserves or nearby mines. In the early nineteenth century, town gas spread as a source of illumination across Europe and the US, and then Japan. In these days, gas was very much a local business. When a substantial oil industry developed in the US, natural gas was often found alongside oil, but the gas was mostly flared (i.e., burned in a controlled manner). While the first pipelines emerged in Texas and Oklahoma in the 1890s, it was only after World War II that a significant network of gas pipelines would be built across the US.[57] The period 1945–70 marked a golden age for gas in the United States, with production, pipeline construction, and consumption rising quickly. By the late 1960s, natural gas provided one-third of total primary energy consumption in the US.[58]

Before 1960, the development of the natural gas industry was essentially a US phenomenon. In Western Europe, "manufactured" town gas would continue to dominate until the 1960s.[59] Italy made the first move to gas after it had found natural gas in the Po Valley during World War II. But it was really the discovery of natural gas in and around the North Sea that would make this region the center of Western Europe's natural gas industry. In particular, the discovery of the Groningen gas field in the Netherlands in 1959 set the stage for rapid growth there, and also in neighboring Belgium, Germany, and France in the 1970s.[60] The UK decided to switch from coal to natural gas following the discovery of the West Sole gas field in the southern North Sea in 1965.[61] The late 1950s was also the era when the gas-rich Soviet Union adopted an industrial strategy that mandated a shift from coal to gas.[62] The epicenter of the Soviet gas industry were the gas fields in Ukraine, which were discovered before World War II, and, in the 1960s, major gas fields were discovered in Western Siberia, including the Urengoy field, at the time the world's largest.[63]

The 1973 oil shock gave further impetus to a shift to natural gas in Europe and Japan, as a means to diversify away from oil. In the following decades, consumption of natural gas rose rapidly in the (Former) Soviet Union, Western Europe, and later also Asia (mostly Japan and South Korea). As result, North America's share of global gas consumption fell from close to 90 percent in 1950 to around 25 percent in 2017.[64]

The development of long-distance pipelines

It took a while before substantial volumes of natural gas were traded across borders. By 1960, the only significant export stream was gas from West Canada going to the US Midwest,[65] but Dutch exports to Europe would commence as from the late 1960s. In 1970, less than 5 percent of globally produced gas was traded internationally. By 2017, this share had gone up to almost 33 percent (figure 2.5). Most of the international trade in natural gas is still pipeline bound, but the share of LNG has been growing steadily.

The most extensive pipeline network has been built in Eurasia. The Soviet Union began exporting gas to Czechoslovakia, a Warsaw Pact ally, in 1967 through a new pipeline which was called, one year before the Soviet invasion, "Bratstvo" (Brotherhood). The initial trajectory of Brotherhood was from the Shebelinka gas field in Eastern Ukraine to the city of Uzhgorod, and then into Czechoslovakia. In the following years, the Brotherhood system would be extended and develop into the main artery to bring Russian gas into Europe. In 1968, Austria was the first country outside of the Soviet bloc to import natural gas from the USSR via a short spur from the Brotherhood pipeline. West Germany followed in 1973, Italy in

BCM

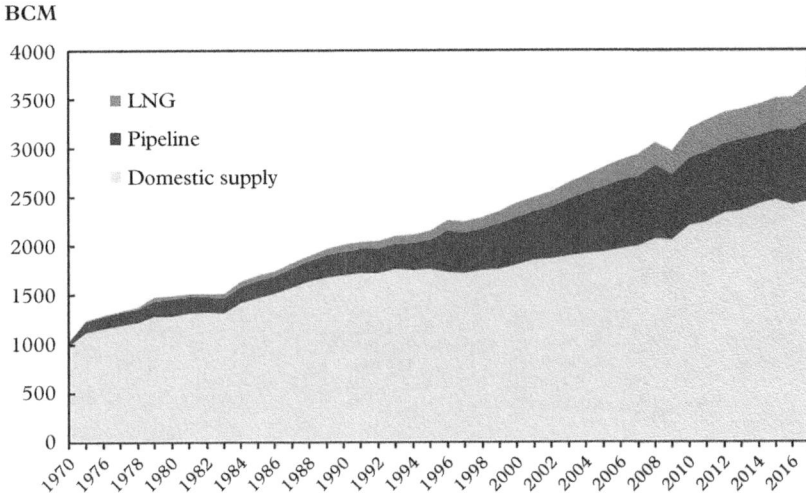

Figure 2.5 Evolution of international gas trade, 1970–2017

Source: Compiled with data from CEDIGAZ, Country Indicators (September 2018).

1974, and France in 1976. By that time, the Soviets had built another major pipeline, called "Northern Lights," from the giant Urengoy gas field in Western Siberia to Minsk in Belarus and then to Uzhgorod in Ukraine, where it fed into the Brotherhood transmission system.[66] This pipeline also allowed tie-ins to the Baltic states and deliveries to Finland began in 1974. By 1978, another pipeline was completed: "Soyuz" (Union). It ran from Orenburg (just north of Kazakhstan) to the Uzhgorod export hub on the Ukraine–Slovak border.[67] Figure 2.6 shows the major East–West gas export pipelines.

This crucial formative period of the East–West gas trading system, the period 1965–75, fell in the midst of the Cold War. The flow of natural gas across the Iron Curtain was seen as a vehicle to improve East–West relations. For Willy Brandt, who became the Foreign Minister (in 1966) and then Chancellor (in 1969) of West Germany, it was a crucial part of his *Ostpolitik*, his policy of engagement with the Soviet Union.[68] On the Soviet side, the Ministry of Gas Industry (Mingazprom) saw in the exports a way to expand the *domestic* role of natural gas, and it was also eager to buy high-quality compressors, steel pipes, and communications equipment available in Germany, Italy, and the UK.[69]

It was only in the early 1980s that Soviet gas exports to Europe would become subject to major political debates. Western European

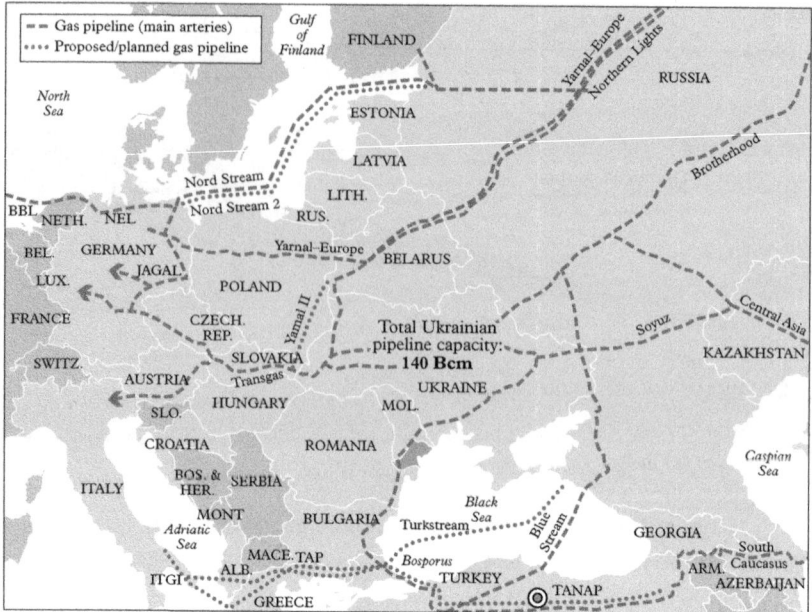

Figure 2.6 Europe's major gas pipelines ties to Russia

Source: S&P Global Platts, © 2019 by S&P Global Inc., reproduced with permission.

countries had reached an agreement with the Soviets to import larger volumes of gas from Western Siberia and build a new pipeline for that purpose. However, the newly elected US president Ronald Reagan opposed the project, which he regarded as a threat to Europe's security. In particular, he feared that the Soviet Union might use (the threat of) gas supply disruptions for political blackmail against Western Europe's NATO members. The US tried to stop the project by placing an embargo on sales of pipeline-related technology and equipment to the USSR and European companies, but its efforts were in vain.[70] In late 1983, the Urengoy–Uzhgorod pipeline was ceremoniously inaugurated.[71] It basically extended the Brotherhood system to Western Siberia. For the first time, but not for the last, Europe's gas dependency on Russian gas had raised geopolitical concerns.

The dissolution of the Soviet Union led to the emergence of transit countries along the East–West supply route. In 1992, 90 percent of Soviet gas exports passed through Ukraine.[72] Ever since, alternative pipeline routes have been built that circumvent Ukraine.

They include: the Yamal-Europe pipeline, in operation since 1999, which runs along the Northern Lights pipeline until Belarus, where it branches off into Poland; the Blue Stream pipeline, in operation since 2003, which goes through the Black Sea to Turkey; and the Nord Stream pipeline, in operation since 2012, which goes directly to Germany through the Baltic Sea (see figure 2.6).[73]

LNG, shale, and the globalization of gas markets

Gas is increasingly traded on tankers just like oil. To do so, the gas needs to be cooled to $-162°C$ and stored in liquid form. At that moment, it becomes liquefied natural gas (LNG). LNG's key advantage is that it takes up only 1/600th of the volume of gas in its natural form. The technology for transporting natural gas in LNG tankers has been around for a long time. The first LNG cargo was shipped from the US to the UK in 1959, on an experimental tanker called the *Methane Pioneer*. It was followed in 1964 by the first commercial exports of LNG from Algeria to the UK and France. By 1969, Libya was also exporting LNG to Spain and Italy, while the US was exporting LNG from Alaska to Japan.[74]

Yet, it is only in the last couple of decades that LNG has started to play a significant international role. After 1975, LNG trade would develop mostly in the Asia-Pacific basin, where the major suppliers are Indonesia, Malaysia, Australia, and Brunei, and the major destinations are Japan, Korea, and Taiwan. Today, LNG accounts for around one-third of global trade in natural gas, and about 10 percent of total gas consumed annually. As of 2018, there were 19 exporters of LNG and 37 importers. Qatar has emerged as the world's largest exporter of LNG, with a global market share of 25 percent.[75] The Asia-Pacific basin still dominates LNG trade, yet the center of gravity is expected to shift more to the Atlantic basin, as a result of the US shale revolution.

The American shale revolution started to unfold in the period 2003–5. Shale gas is methane that is trapped in tight layers of sedimentary rock of low permeability. The breakthrough in shale extraction was enabled by the combination of three technologies: hydraulic fracturing (or, simply, "fracking"), horizontal drilling, and seismic imaging. While the technique of fracking is not without controversy (see box 2.2), it unlocked vast swaths of natural gas deposits in the US. Large-scale shale gas production began in 2006 at the Barnett formation in Texas, but soon spread to the Haynesville fields on the border of Louisiana and Texas, and the giant Marcellus field, which

---

Box 2.2   The controversial technique of shale gas hydraulic "fracking"

Hydraulic fracturing or fracking is a process in which a fluid is pumped under high pressure into tight, underground rocks. This high-pressure injection fractures the rock and releases the oil or gas resources that were trapped inside of it. The fluid consists of mainly water and sand but contains chemical additives designed to ensure that the sand penetrates the cracks and keeps them open, while the oil or gas molecules flow to the surface. The technology has been known for decades, primarily as a mechanism to release oil from elderly and declining wells.

Fracking has become controversial, however, as the success of the documentary *Gasland* (2010) has illustrated. It raises two sets of concerns. The first set of issues revolves around water. Fracking is an industrial process that uses significant amounts of water. The waste water will contain the original chemicals but will also bring with it trace quantities of heavy metals, benzene and various other carcinogens. The waste water has to be treated before it can be released or it can contaminate rivers and potentially even underground aquifers.

A second set of concerns has to do with climate change. There is no doubt that fracking operations in the US have caused serious amounts of methane, a powerful greenhouse gas, to leak into the atmosphere. If you take into account these methane leaks at the well, shale gas could be even worse for the climate than coal when used in power plants.[76] Thus, this concern about fracking is that it unlocks a large new source of fossil fuels at a time when we need to reduce carbon emissions.

---

cuts across five US states, including Pennsylvania and New York. Shale gas production went from zero in 2000 to half of total US natural gas production in 2015 – a game changer for US markets. It contributed to a drop in gas prices and a decline in coal-fired electricity generation in the US, along with an increase in US coal exports.

The shale boom has created global ripple effects. Whereas the US was preparing itself for more gas import dependence in the first half of the 2000s, by 2009 it had surpassed Russia as the world's largest gas producer. Plans for building LNG import terminals along the US coastlines were reconfigured into plans for *export* terminals. The first cargo of LNG left from a port in Louisiana in February 2016. Since then, more than a dozen liquefaction units at six export terminals have been commissioned or approved.[77] At the same time, LNG suppliers that had eyed the US as a guaranteed market had to find other markets.

So far, shale gas production has been confined largely to North America, but recoverable shale gas resources are to be found in many countries across the world. In fact, China, Argentina, and Algeria are thought to have larger shale gas reserves than the US.[78] Other countries have tried to replicate the US shale success, notably in China and Europe. China has made some progress but, in Europe, high initial expectations for a shale gas boom have not been met with reality.[79] Currently, the United Kingdom, where exploratory fracking has begun in Lancashire, is the only country still moving forward.[80] Yet, the US has a number of unique features that make it difficult to simply copy-paste its experience to other countries: its shale has high concentrations of hydrocarbons and is brittle enough to allow fracturing; subsurface mineral rights are owned by landowners; it has favorable taxation and legal regimes; and water, sand, pipelines, and capital are readily available.[81] A number of US states have banned fracking, as have countries such as France and Bulgaria.

The upsurge in shale production together with the growth in LNG trade could facilitate the emergence of a truly global gas market much like the one we have for oil. The pricing and contract structure of the international gas business are already undergoing major changes. Traditionally, the business model for pipeline gas and LNG involved long-term contracts of 20–30 years. They included "take-or-pay" requirements, committing buyers to pay for a minimum annual volume of gas regardless of whether or not they actually purchased the entire volume of contracted gas. Under those types of contracts, the price of gas was linked to the price of oil which, until the 1990s, was considered a competitor for gas.[82] This market structure allowed the industry to earn back its high upfront capital investments in the expensive long-distance pipelines and LNG terminals that needed to be constructed before any international gas trade could take place.

Indexation to oil prices has begun to erode as hubs – trade platforms where natural gas can be traded virtually or physically – started to emerge in the US, making their way to northwestern Europe via the UK. The Henry Hub is America's best-known hub, while the National Balancing Point (NBP) in the UK and Zeebrugge in Belgium are two major trading hubs in Europe. Thanks in part to the proliferation of these hubs, more and more gas is traded on the spot market, where prices are determined competitively, rather than through long-term contracts. Although long-term, oil-indexed contracts may remain prevalent in some markets, it is clear that gas is increasingly traded as a global commodity.[83]

The implications of this were vividly demonstrated in March 2011. In the wake of the earthquake and Fukushima nuclear accident, Japan had to close all of its 50 nuclear power stations, leaving the country with no nuclear generation from September 2013 to August 2015. Existing coal-fired power plants were operating near full capacity, so utilities had to import large volumes of LNG to meet electricity demand. In the old, inflexible and regionally based LNG industry, it would have been impossible to redirect gas flows on such a massive scale to Japan. Yet, luckily for Japan, natural gas had become a global commodity.

Other fuel markets: Coal and uranium

Even though oil and gas are vitally important to the global economy, they are not the only significant energy markets. At least two others deserve mention: coal and uranium.

Coal

The modern industrialized world was built to a large extent with coal. The Industrial Revolution which started in Britain in the late eighteenth century and spread to the rest of Europe, to North America, and beyond, was largely coal-fueled. Since then, coal's use has only grown, although it has stagnated since 2012 (see figure 2.7).[84] Coal is still the backbone of the world's electricity sector, and the second largest primary energy source overall. In around 2000, coal consumption in developing countries started to increase in an unprecedented way. Yet, we may be nearing an inflection point. China's massive coal expansion seems over, coal is in terminal decline in North America, and the appetite for coal in many other parts of the world has cooled due to coal's contribution to air pollution and climate change.[85]

Contrary to the globalist nature of oil and natural gas, the story of coal is largely a domestic one. Since the biggest consumers of coal also have very large reserves, international trade in coal has remained limited. The proportion of coal that was traded across borders historically stabilized at around 15 percent,[86] a level at which it remains today. This percentage is much lower than those for oil and gas. Even

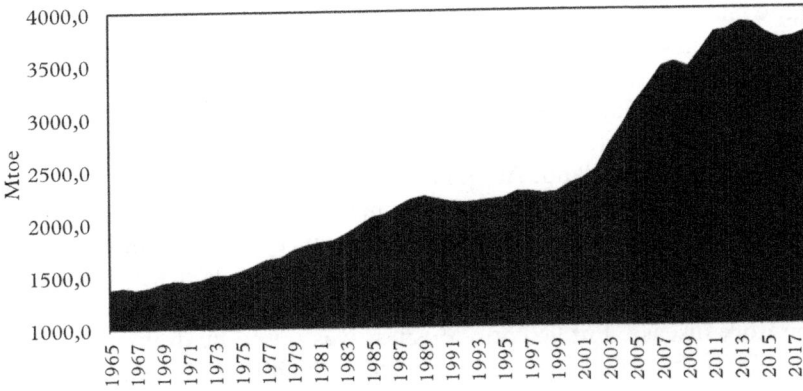

Figure 2.7 World coal consumption, 1965–2018

Source: BP (2019) *Statistical Review of World Energy.*

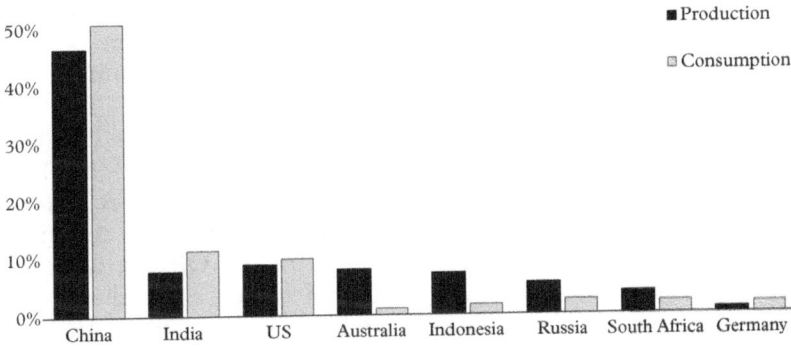

Figure 2.8 Key players in global coal consumption and production (2017)

Source: BP (2019) *Statistical Review of World Energy.*

so, the international dimension colors the policy choices regarding coal: coal expansion has often been promoted for energy security reasons. After the oil turbulence of the 1970s, for instance, the OECD countries would increasingly displace oil with coal to generate electricity.[87] Given the abundance and geographic dispersion of coal reserves, no one has ever appeared particularly concerned about the emergence of a "coal cartel" or the use of a "coal weapon."

The story of coal today is also largely a Chinese one. It accounts for half of the world's consumption of coal, and it is also the largest producer, accounting for 46 percent, followed by India and the US (see figure 2.8). In the 2016–20 Five-Year Plan, China for the first time laid out the goal of capping coal consumption in absolute terms and reducing its share of primary energy to 58 percent by 2020, down from 64 percent in 2015.[88] Japan is the world's largest coal importer, followed by India, China, South Korea, and Taiwan. Australia is the world's biggest coal exporter, and even has a bigger market share in coal exports than Saudi Arabia has in oil exports.

Coal's future is constrained by its negative environmental impacts. Generally speaking, coal releases almost twice as much carbon dioxide when combusted as natural gas. It also is a key source of outdoor air pollution, which contributes to health problems from respiratory disease. Public consciousness was raised by dramatic events such as London's killer smog of 1952, which directly caused 4,000 premature deaths and led the UK parliament to ban the use of soft coal in cities.[89] From the 1980s, Europe and North America would also take steps to reduce sulfur emissions from coal-fired power plants, in an effort to tackle the issue of acid rain. More recently, in response to the so-called "airpocalypse" of 2012–13 in Beijing, China has taken steps to curb coal consumption and speed up the use of filters and scrubbers.[90]

At the climate conference in Bonn, in late 2017, the UK and Canada launched the Powering Past Coal Alliance (PPCA), a voluntary partnership of state and sub-state governments, business actors, and other organizations who want to phase out unabated coal-fired electricity generation no later than 2030 in the OECD, with the rest of the world no later than 2050.[91] However, phasing out coal will not be easy. Coal mining is much more labor-intensive than oil and gas extraction, and closing down coal mines can create a lot of unemployment in certain regions. Moreover, coal is still regarded as a cheap source of energy that can provide baseload electricity supply. Coal's price is low because it does not account for the cost of negative externalities (air pollution, climate change, black lung disease with coal miners), but also because the sector is heavily subsidized globally (much like the fossil fuel sector in general).

Nuclear and uranium

After its emergence in the 1940s as part of the Manhattan Project and the rush to build nuclear weapons during World War II, nuclear power saw a rapid expansion in the 1970s and 1980s. Yet, over the past three decades, the growth of nuclear power has slowed down, with the share of nuclear in electricity generation actually declining from 17 percent in 2000 to 10 percent in 2017.[92] There are currently

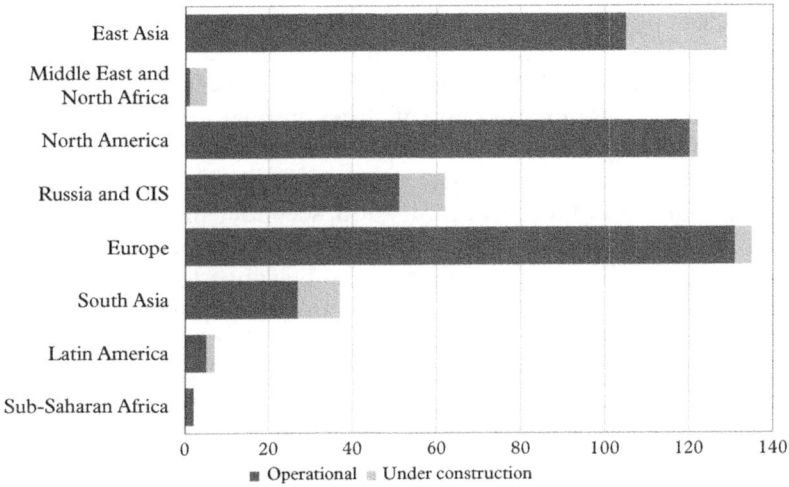

Figure 2.9 Regional distribution of nuclear power plants

Source: Compiled with data from IAEA (2018) *Nuclear Power Reactors in the World.*

454 nuclear power reactors in operation, spread over 30 countries (figure 2.9).[93] Around two-thirds of today's nuclear fleet in advanced economies is more than 30 years old, and will be shut down in the foreseeable future unless their lifetime is extended.[94] Some countries are building new nuclear power plants, notably China, India, Russia, and the UAE. Others are planning to phase out nuclear power, including Germany, Switzerland, Spain, and South Korea.

The growth prospects for nuclear energy are highly uncertain. On the one hand, nuclear energy – with its high energy density and low carbon footprint – remains attractive to some countries in terms of contributing to decarbonization and energy security. On the other hand, nuclear power plants rely on non-renewable resources like uranium; produce toxic radioactive waste; can pose a threat to public health and the environment in the case of accidents such as those in Chernobyl (1986) and Fukushima Daiichi (2011); raise concern about nuclear proliferation; and are becoming ever more complex and expensive to build, operate, and decommission.

The key primary fuel for the current generation of nuclear fission reactors is uranium, a naturally occurring element in the Earth's crust. Uranium mines operate in some 20 countries, although half of world production comes from just ten mines in six countries: Canada, Australia, Niger, Kazakhstan, Russia, and Namibia.[95] It is interesting to note that many countries with large uranium reserves do not rely on nuclear power, like Australia and Kazakhstan. In fact, the only two countries that mine enough uranium to meet their domestic needs are Canada and South Africa, but even these two rely on other countries for uranium enrichment. International uranium trade has never led to major security or supply concerns like those associated with oil and gas because, first, uranium production has exceeded demand for most of the last century and, second, a high proportion of uranium resources comes from stable Western democracies.[96]

Importantly, however, uranium has to be converted, enriched and fabricated into a fuel before it can be of any use for owners of nuclear power plants. After the fuel is used, it can either be reprocessed or stored. These different activities – conversion, enrichment, fuel fabrication, reprocessing, and waste management – are generally referred to as the "nuclear fuel cycle." Only a handful of countries have managed to build up the full nuclear fuel cycle capacity, namely the US, UK, France, China, and Russia. These countries therefore play a key role in the international market for uranium. Yet, the market for uranium does not function like other commodity markets. There is no elaborated spot market or futures trading. Instead, uranium trade relies primarily on bilateral deals, often long term in nature, between a producer and a consumer.[97]

Only four frontrunner states – the US, Canada, the UK, and Russia – have managed to build their nuclear programs independently, based on domestic technology. For all other countries, installing nuclear power has introduced a new set of dependencies, whether it is dependence on the import of reactors, fuel, or expertise. Some countries strike so-called "build-own-operate" agreements, whereby foreign firms will build nuclear power plants in exchange for ownership and a guaranteed electricity price.[98] Since atomic energy is in steady decline in the OECD, Russia has positioned itself as the key supplier of nuclear reactors and technology to the growth markets in the developing world. This has raised some fears in the US about Russia's growing nuclear dominance, which could elevate Russia's influence and leverage on the geopolitical scene.[99]

International trade aspects of renewable energy

Renewable energy was once referred to as alternative energy, too expensive to expand beyond niche markets. Thanks to the extraordinary growth in renewables over the last couple of years, that perception has now changed. Since 2012, renewables have added more capacity to the global electricity sector than fossil fuels and nuclear combined. Consequently, a third of global power capacity is now based on renewable energy.[100] Most of the growth in renewables has been confined to the power sector but, thanks to the ongoing electrification of our economies, they will also find their way to other end-use sectors. Electric vehicles and heat pumps, for example, are extending the deployment of renewables in sectors like transport, industry, and buildings.

Just like the term "fossil fuels," the term "renewables" covers a wide basket of different energy sources. The main renewable energy sources are bioenergy, geothermal, hydropower, ocean, solar, and wind. Among these, solar energy and wind power are undergoing very rapid growth, while the others are growing more gradually. Solar and wind share a characteristic that is largely unique to them: the amount of power they generate varies with the weather and the time of day. This is why they are called "variable" renewable energy sources.

The growth of renewables is driven by technological innovation and declining costs, increasingly making renewables competitive with conventional energy sources. Hydropower, geothermal, and biomass are mature technologies that have long been a source of low-cost electricity where they have been built. The cost of electricity from photovoltaic (PV) solar panels fell by 77 percent in 2010–18, and the cost of onshore wind by 34 percent over the same period. The International Renewable Energy Agency (IRENA) expects all types of renewable energy to be cost competitive with fossil fuels in most places in the world by 2020 (figure 2.10).[101]

Renewable energy has its own geography of trade. Generally speaking, renewables like solar and wind represent a shift from fuel markets to hardware markets. In contrast to fossil fuels, solar or wind energy are not traded directly. Instead, the goods and technologies to harvest these energy flows are traded. This includes everything from solar PV panels over smart meters to batteries, but also their components and parts (for example blades for wind turbines or water wheels for hydropower) and related services (for example engineering and installation services).[102]

The electricity generated from renewables (or from other sources) can also be traded across borders. Unlike oil or LNG, however, electricity is not a globally traded commodity. The reason is that much of the electricity is lost when transported over long distances with current technologies. Europe is a region where strong electricity interconnections have been built, particularly in the Nordic area. Yet, regional electricity networks are being constructed in many parts

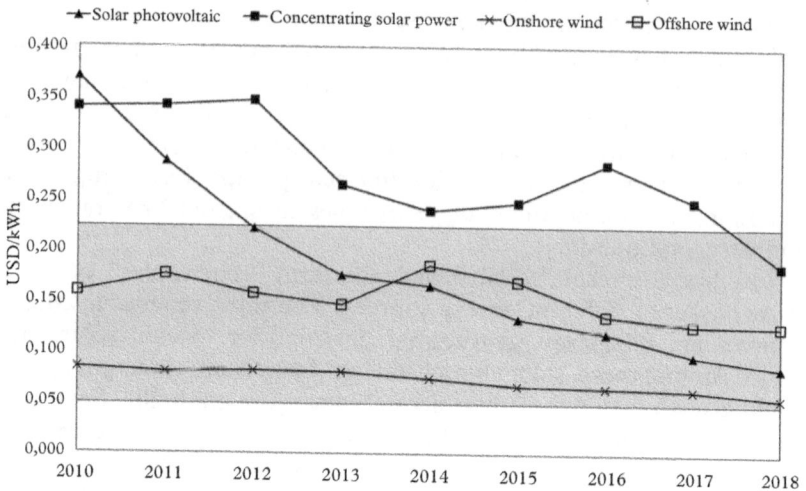

Figure 2.10   Renewable power generation costs, 2010–2018

Note: Grey area denotes fossil fuel power generation cost range.

Source: IRENA (2019) *Renewable Power Generation Costs in 2018.*

of the world, including Latin America, Africa, and the Persian Gulf region. Interconnections make grids more stable and resilient, which is very useful when you integrate higher shares of variable renewable energy sources.[103] In recent years, several renewable energy supergrids have been proposed, including Desertec (to bring renewables from North Africa to Europe), Gobitec (to bring renewables from Mongolia to major urban centers in Asia including Seoul and Tokyo), and the North Sea Offshore grid. What is more, State Grid, China's state-owned electric utility company, has plans to develop a global supergrid called the "Global Energy Interconnection" that will link every continent with undersea transmission cables to power the world with green electricity.[104]

One type of renewables where there is a global commodity market is biofuels. In contrast to wind, solar, and hydro, the primary use of biofuels is not for generating electricity but in transport, to power cars and planes. Biofuels have become significant commodities in world markets, particularly since 2006. Ethanol has been traded globally in significant volumes for decades. Biodiesel trade is less established and continues to consist primarily of trade in feedstocks (such as soya beans and vegetable oil) rather than fuel. The triangular trade relationship between the EU, Brazil, and the US is still the backbone of international biofuel trading, supported by blending mandates in those markets.[105] Biofuels are subject to spot and futures trading on international commodity exchanges such as the one based in London.

In the future, other renewable fuels – that is liquid fuels generated from renewable energy sources – could be traded in large quantities. Areas that possess an abundant supply of renewable energy, such as Patagonia or the Australian desert, might begin large-scale production of renewable fuels such as hydrogen, ammonia, methane, and methanol. Such fuels can allow for seasonal storage of renewable electricity (which only pumped hydro has been able to do to date), and make use of existing infrastructure (such as natural gas pipelines). They also have the potential to reduce emissions in hard-to-electrify sectors such as aviation and some industrial processes.[106]

Conclusion

This chapter sought to provide an overview of the history and functioning of key energy sources. Oil attracted most of the attention as it was the first energy source around which a large international market was established. The history of oil is intimately connected to key events in international affairs more broadly, such as the two world wars, decolonization, and the collapse of the Soviet Union. No other energy sector is as internationalized as oil. The industries that first developed around coal, electricity, and nuclear were initially nationally segregated industries, while natural gas started off as mostly a local business. Over the past couple of decades, as global energy demand has vastly expanded, many of these sectors have begun to go global, with natural gas being the prime example here. As modern renewables can be deployed at any scale, from the household to the utility level, they may decentralize the global energy system again, but the global value chains will continue to exist.

Now that we have established a basic understanding of the key energy resources and markets, the following chapters will look at their impact on geopolitics, economics, the environment, and global justice.

PART I

World politics through an energy prism

3

Energy and security:
Fueling geopolitics and war?

Since the Industrial Revolution, fossil fuels have been a key driver of the geopolitical landscape. The world's highly concentrated deposits of oil, natural gas, and coal have helped shape patterns of power and wealth, and hence cooperation and conflict, in the international system. Access to high-yield energy sources has been a key currency of power for states and lack thereof a sign of vulnerability. Coal and steam power were essential pillars of the British Empire in the nineteenth century whereas control over oil production and trade formed the bedrock of US power in the twentieth century.[1] The ongoing transition to renewable energy could be as transformative for the global balance of power as the historical shifts from wood to coal and from coal to oil.[2]

A state's position in the international system is obviously influenced by many attributes, including population, land size, and natural resources. Oil wealth often translates into military strength as the massive influx of petrodollars allows countries like Saudi Arabia to have enormous defense budgets and buy state-of-the-art weaponry. It also allows them to accrue huge dollar reserves which they can then "recycle" – that is, spend it to finance domestic consumption and investment or saving it in foreign assets (often US securities). The fact that the dollar is used to settle most international oil trade offers a huge geostrategic advantage to the US because it has allowed it to continue to borrow at unrealistically low interest rates.[3]

Conversely, lack of access to energy can be a source of geostrategic vulnerability and creates concerns over "energy security" – that is, the challenge of assuring available, affordable, and reliable flows of energy. In their "scramble" for energy resources, for example, Chinese firms have embarked upon an "oil safari" with billions of dollars of oil contracts signed with Angola, Ethiopia, the Congo, Gabon, Kenya, Madagascar, Nigeria, and the Sudan. These firms have followed a "no-questions-asked" policy of "non-interference," meaning they "don't hold meetings about environmental impact assessments, human rights, bad governance and good governance."[4] In Sudan, the Chinese National Petroleum Corporation funneled $4 billion to the regime in Khartoum, enabling them to purchase Shenyang fighter planes and an assortment of heavy arms that they then used to try and suppress a secessionist movement in the south.

As these examples show, control over energy resources is a key determinant of power, wealth, and, indeed, conflict. Focusing on these hard-nosed aspects of international energy trade corresponds to the dominant narrative or "frame" of global energy politics, that of geopolitics, which is akin to the realist school of thought in international relations. In this frame, energy is seen as a currency of power, a strategic tool in foreign policy, and a key source of conflict.[5] This chapter reviews these three aspects, before taking a closer look at two specific areas – nuclear proliferation and the geopolitics of the global energy transition.

Energy statecraft

Some authors claim that decades of progressing globalization have brought us to an age of "connectivity" that creates opportunities for states to "weaponize interdependence,"[6] and to pursue war by other, predominantly economic, means.[7] For students of global energy politics, this sounds familiar. States have long used – or tried to use – energy resources and related technologies as instruments of foreign policy, a practice known as "energy statecraft." There are at least three aspects of international energy trade and investment that states can attempt to "manipulate" for geopolitical purposes: energy flows, prices, and infrastructure.[8] These different techniques of energy statecraft can serve goals that are benign – for example, to foster peace and interdependence between countries – or less benign – to exert geopolitical leverage over other countries.

Manipulating flows

States can attempt to manipulate cross-border flows of energy directly. They can do so through sanctions or boycotts, as in the iconic case of the 1973 Arab oil embargo. This politically motivated boycott probably affected no country more than Japan – 77 percent of its energy consumption was oil, and most of it came from the Middle East. Since the end of World War II, Japan's foreign policy had centered on an alliance with the United States. Henry Kissinger, the newly appointed US Foreign Secretary, flew to Tokyo in November 1973 to persuade the Japanese government not to capitulate to the Arab demands. Just a few days after Kissinger's visit, however, Japan did just that. It publicly declared that it endorsed the Arab position in the conflict against Israel.[9] This move "was the first open break with American foreign policy in postwar diplomatic history that Japan had dared to make."[10]

In the case of natural gas, the much-publicized Russian–Ukrainian gas crises have often been interpreted as instances of the so-called Russian "gas weapon." Gas deliveries from Russia to Ukraine were indeed cut off in January 2006, in January 2009, and in June 2014. The fact that the state-controlled company Gazprom has a legal monopoly over gas export flows only reinforces the view that the Kremlin can decide to open or close the gas taps for geopolitical motives. It has led to calls by European politicians to "disarm Russia's energy weapon."[11] While there is little doubt that Russia has on occasion tried to use its position as a major gas supplier to countries in its "near abroad" as a tool to gain political leverage, the example is less explicit than the Arab oil embargo.[12] Russia's gas threats have always been cloaked in a discourse of commercial and legal obligations.[13]

Another opportunity to disrupt physical flows of energy lies not at the upstream, but at the midstream segment – that is, transportation. Grid-based energy flows like pipelines or electricity transmission lines offer opportunities for sabotage or attacks. Long-distance pipelines that cross multiple countries can be disrupted by the transit countries.[14] While most of the time, these pipelines operate without any notable problem, politically-motivated disruptions have occurred.[15] A case in point are the gas exports from Turkmenistan to Europe via Russia. In the mid-1990s, Gazprom, as the sole owner of the pipelines that took Turkmen gas to Europe via Russia, halted Turkmen exports and demanded an increase in transit fees. Turkmenistan had little choice but to agree, as there were no viable alternative export options at the time.[16]

For oil, transportation risks are mostly associated with the so-called "chokepoints" – that is, narrow waterways through which a significant number of oil tankers pass every day (see figure 3.1). Very recent seizures of international oil tankers by Iran in the summer of

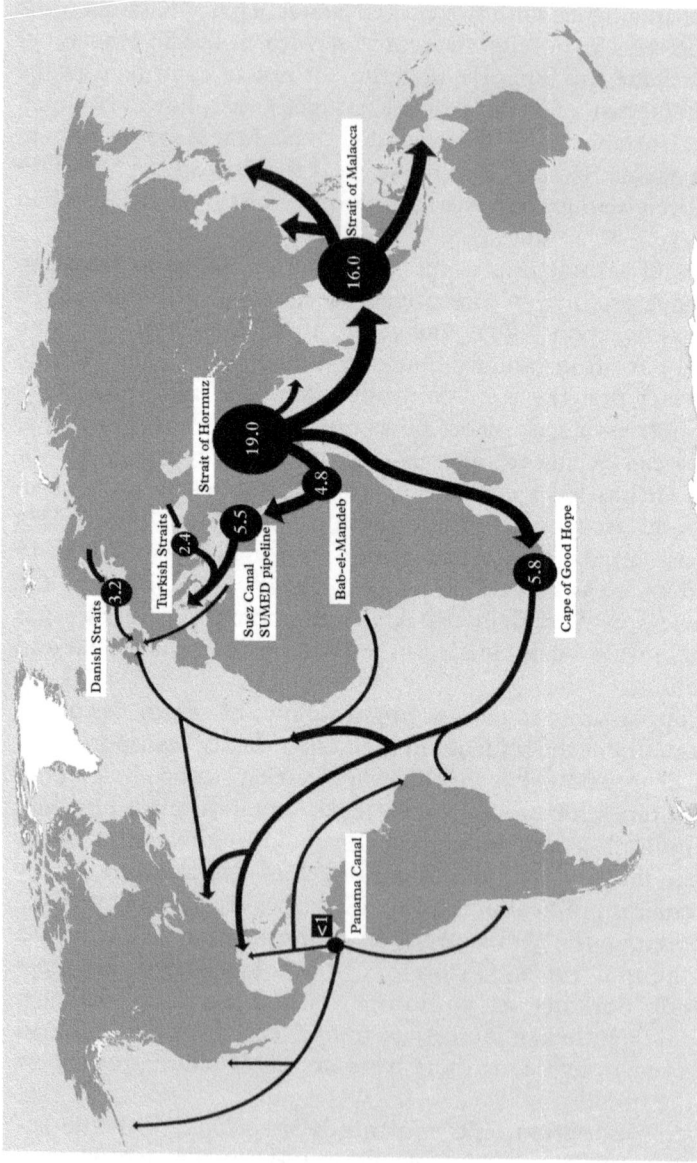

Figure 3.1 Major maritime oil flows and chokepoints

Note: All estimates in million barrels per day. Includes crude oil and petroleum liquids. Based on 2016 data.

Source: IRENA (2019) *A New World: The Geopolitics of the Energy Transformation*, based on data from the EIA. Reproduced with permission. Accessible at https://www.irena.org/publications/2019/Jan/A-New-World-The-Geopolitics-of-the-Energy-Transformation.

2019 signify the strategic importance of chokepoints.[17] The Strait of Hormuz, which links Middle East crude producers to key global markets, is the world's most important oil artery. At its narrowest point, it is only 21 miles wide and the shipping lane is only 2 miles wide in either direction. Each day, around 30 percent of all seaborne traded crude oil passes through it, as well as a significant volume of LNG. There have been several military incidents in the Strait, such as the "tanker wars" between Iran and Iraq in the 1980s.[18] Iran has threatened several times to disrupt the flow of energy through Hormuz, often in response to sanctions by the United States.[19] In the summer of 2019, several oil tankers were seized by Iran in the narrow Strait, leading several Western countries to send additional warships to the region. In Beijing, strategic planners are more concerned with the US naval dominance in the Persian Gulf, which it could exploit to disturb oil exports to the East in case of conflict between the US and China.[20]

It is not just the exporters or transit countries who have wielded the "energy weapon." Importers, too, may make use of the embargo tool. In the mid-1980s, for example, Europe and the United States joined the efforts of a number of developing countries to place an oil embargo against Apartheid South Africa.[21] Another high-profile case is the oil sanctions against Iran (see box 3.1). These sanctions are enacted unilaterally and thus differ greatly from the United Nations (UN) oil sanctions imposed on Saddam Hussein's Iraq between 1990 and 2003, amended in 1996 to allow for limited oil exports in return for food imports (the so-called "Oil-for-Food Program").

Some sanctions do not curtail a country's current exports, because doing so would raise global prices and inflict equal pain on those imposing the sanctions. Instead, sanctions are targeted to hinder investment and trade, thereby affecting the future growth of the energy industry. This approach underpinned the oil sanctions that Western states and Japan imposed on Russia in 2014, after its annexation of Crimea. Blocking Russia's exports, which hovered above 10 mb/d at that time, would have been impractical and dangerous to the global economy. Equally so, blocking Russia's 160 billion cubic meters in annual gas exports to Europe would cripple the European economy as much as it would cripple Russia's. Therefore, the post-2014 energy sanctions against Russia only cover oil, not gas, and they are designed to constrain future production growth from Russia's Arctic, shale, and deepwater oil reserves, rather than to curb current oil production.[22]

Countries have sometimes tried to exert geopolitical influence,

---

Box 3.1   History of the oil sanctions against Iran

In response to the Iran hostage crisis of 1979, the United States banned Iranian oil imports. Quite soon, it learned that its oil embargo was riddled with gaps and loopholes that made it ineffective. Because oil is a fungible commodity, Iran's oil could simply be swapped with other countries' oil for import into the United States, while Iranian oil easily found its way to other export markets.[23]

In the mid-1990s, the United States therefore moved from applying strictly unilateral oil sanctions to an attempt to enforce extra-territorial sanctions through the adoption in 1996 of the Iran-Libya Sanctions Act (ILSA).[24] Under this Act, the US threatened to impose sanctions on foreign firms doing business with Iran, yet the full strength of the Act was never applied to the foreign companies investing in Iran's energy sector.[25]

In 2012, following revelations that Iran might have worked on nuclear explosive technology, both the European Union and the United States imposed an oil embargo against Iran. At the time, Iran was still the third largest oil exporter in the world. The EU gradually halted all of its imports from the country, while the US put in place financial sanctions against any potential buyer of Iranian oil. These sanctions were quite effective in curbing Iran's oil exports.[26] They arguably helped to pave the way for the 2015 Iran nuclear deal, the so-called Joint Comprehensive Plan of Action.

The oil sanctions against Iran were consequently lifted but, in 2018, US President Trump withdrew from the Iran nuclear deal and re-imposed oil-related financial sanctions.

---

not by withholding supplies of oil and gas, but by doing exactly the opposite: by flooding the market. After oil prices fell dramatically in 2014, for instance, OPEC did not move to cut production because it preferred to maintain its market share in the face of growing US shale production. Whether this move should be interpreted as a business strategy linked to profitability or as a strategic foreign policy move, for instance to inflict harm on Saudi Arabia's geopolitical rival Iran, is unclear. In 1986, as mentioned in chapter 2, Saudi Arabia had already opened the oil taps to regain market share it had lost in part to its fellow OPEC members, many of whom had failed to abide by their quota.

It is not just oil and gas flows that can be manipulated for political purposes but other resources too. Uranium exports to India have been banned since the entry into force in 1970 of the Non-Proliferation Treaty (NPT), which was not ratified by India. Despite the ban, India successfully tested its first atomic bomb in 1974. Eventually, after the 2008 US–India nuclear deal, most nuclear trade sanctions were lifted.[27] Coal exports from North Korea have been completely banned by a United Nations Security Council resolution from 2017.[28] Rare earth minerals, which are critical for several renewable energy technologies, experienced a major price hike in 2010 when China restricted the supply to foreign buyers after a naval stand-off with Japan. Beijing firmly rejected accusations that it was manipulating the supply of rare earths for political purposes, but the Chinese export restrictions were successfully challenged by Japan, the EU, and the US at the World Trade Organization (WTO).[29]

Manipulating prices

Instead of manipulating physical flows of energy directly, governments can also try to influence them indirectly, by manipulating prices. As we have seen in chapter 2, the major oil companies that dominated international oil trade until the 1960s used a "posted price" as a basis to calculate fiscal revenues for oil-producing countries, yet the posted price did not reflect market fundamentals. In the early 1970s, oil producers were able to wrench control over the oil price from the majors. Between October and December 1973, OPEC unilaterally raised the posted price of Arab light crude from $3.65 to $11.651 per barrel. Yet, the OPEC-administered pricing system collapsed in the mid-1980s in favor of a market-based oil pricing system based on liquid spot markets, futures, options, and other derivative markets with 24-hour electronic trading – the latter is referred to as trade in "paper barrels."[30]

Even so, oil countries often still sell crude at discount prices to allies or potential allies. Saudi Arabia and Kuwait have sold oil to Yemen at discount prices, yet many recipients have been fair-weather friends. Yemen supported Iraq after its invasion of Kuwait, despite having received substantial assistance from the Kuwaiti government in the past.[31] In 2010, members of the Commonwealth of Independent States (a club of 10 post-Soviet states) were still able to buy Russian oil at a 35 percent discount, on average, relative to what the rest of the world paid.[32] In Latin America, Venezuela has sold refined oil products on preferential terms to Caribbean nations via the so-called PetroCaribe program since 2005. Yet, due to Venezuela's economic woes, the Maduro regime has had to scale down significantly its contributions to the PetroCaribe scheme.[33]

Natural gas trade lends itself better to price manipulation for political purposes. This is especially the case when natural gas is shipped through pipelines (rather than via LNG tankers). The fixed attributes of cross-border pipelines effectively "lock in" relationships of deep dependency among supplier, transit, and customer states. This creates opportunities for politicization of the gas trade, as the consuming country often has no option – at least not in the short term – to draw on gas supplies from elsewhere. A case in point is Gazprom's natural gas pricing policy toward the former Soviet republics. Russia has certainly used natural gas pricing strategically to buy off allegiances (see box 3.2 for more background).[34] Countries from the former Soviet sphere that leaned toward the West – such as Georgia, Ukraine, and the Baltic states – have been charged with higher gas prices than those who remained friendly to the Kremlin – such as Armenia and Belarus.[35]

Box 3.2   Natural gas pricing and the Russo–Ukrainian "gas wars"

After the break-up of the Soviet Union, some (but not all) Soviet successor states continued to receive Russian gas at discount prices. This changed in the mid-2000s when Russian President Putin began to support Gazprom's desire to realign gas prices for neighboring customers with European oil-indexed prices. The steady increase in the oil prices, and therefore European gas prices, from 2003 to 2008 made the transition particularly difficult for the importers. These price increases provoked a series of "gas wars" between Russia and key transit countries (with Ukraine in January 2006, March 2008, and January 2009; with Belarus in February 2004 and January 2007; and Moldova in January 2006).[36]

Ukraine, especially, was poised to exploit its pivotal transit role for Gazprom's deliveries to Europe.[37] In 2004, 80 percent of Russian gas exports to Europe were still delivered via Ukraine.[38] The 2009 Russo–Ukrainian gas crisis was the most severe. All Russian gas flows to Ukraine were halted for three full weeks in the middle of the winter (not just the gas molecules destined for Ukrainian consumption), with effects reverberating through Central and Eastern Europe. It was resolved when Gazprom and Naftogaz signed an 11-year contract with a price formula similar to those typically used in Gazprom's long-term contracts with its principal European customers. The 2009 contract marked a move away from the previous system whereby the gas price was negotiated bilaterally each year.[39]

In practice, Moscow never fully abandoned its old strategy of buying influence and allegiance through adjusted pricing of its natural gas exports. In April 2010, the contract was renegotiated and Ukraine got a significant discount on the gas price in exchange for a political concession, namely an extension of 25 years on the lease by Russia of the naval base in Crimea. When Kiev was about to sign an association agreement with the EU in November 2013, the Kremlin – through Gazprom – offered a 33 percent discount on the gas import price, from $402 to $268.5 per 1,000 cubic meters.[40] This was probably one of the main factors that lured former President Yanukovych to ditch the trade deal with Europe, triggering the Maidan protest movement that would eventually unseat him three months later.

After the ousting of Yanukovich and the annexation of Crimea, both discounts were scrapped in April 2014. As a result, gas prices for Ukraine hiked by more than 80 percent, to $485 per 1,000 cubic meters, the highest price in Europe. Russia could unilaterally scrap the 2010 discount (obtained for the lease of the Crimean naval base) because the discount was given in the form of a (unilateral) suspension of export duties levied by the Russian state on gas exported to Ukraine. Gazprom was prepared to lower its price demand to $385, broadly in line with prices for other European countries, but Kiev rejected the offer because the discount would have again been awarded in such a way that it could be undone at any time by the Kremlin.[41]

In 2012, the European Commission opened an anti-trust case against Gazprom, accusing it of attempting to partition Central and East European gas markets; hindering customers from reselling gas; charging unfair prices in Bulgaria, the Baltics, and Poland; abusing its dominant market position to control the Yamal transit pipeline through Poland; and bullying Bulgaria to support South Stream. If found guilty, Gazprom could have been fined more than

€10 billion.[42] By May 2018, Gazprom clinched a deal with the European Commission to reform its pricing structure and ban destination clauses (which prevented natural gas from being re-exported by consuming countries).[43] All of the EU's demands had been met. In addition to the legal measures in the gas market, Eastern EU member states also invested in new gas infrastructure – including reverse flow pumping stations, interconnectors, LNG terminals, and storage space. As a result, Gazprom has largely lost its ability to manipulate prices for political purposes, at least toward EU member states.[44]

Manipulating infrastructure

It is not just the energy resources themselves that have been the object of geopolitical bickering, but also the wider energy systems and infrastructure. In the 1990s, for instance, the United States promoted the construction of the Baku–Tbilisi–Ceyhan (BTC) oil pipeline to bolster its strategic influence in the Caspian region at the expense of Russia and Iran.[45] For more than a decade, the EU has supported the construction of a Southern Gas Corridor to reduce its reliance on Russian gas, while both Russia and some European countries have promoted alternative gas routes, such as Nord Stream, to circumvent existing gas delivery routes, which pass through multiple transit states.[46] While the Nord Stream pipeline has been built, plans for the construction of Nord Stream 2, a second gas pipeline from Russia to Germany through the Baltic Sea, have divided Europe. It has also provoked opposition from the United States, who fears the pipeline may compromise Europe's security. In May 2019, a bill was tabled in the US Senate to impose sanctions on companies involved in the Nord Stream 2 pipeline.

Energy-related development aid often plays an important role in shaping energy infrastructures for political motives. During the Cold War, investment in energy infrastructure – such as the World Bank's financing to build large-scale hydroelectric dams in countries such as Vietnam and Ghana – was motivated in part to counter growing Soviet influence.[47] The Soviet Union, for its part, was also happy to provide technical assistance and expertise to a range of developing countries to harness their energy potential – from hydropower in Egypt to natural gas in Afghanistan – and counter the influence of the United States.[48]

Today, Western development aid is being rivalled by Chinese overseas investments. China's two global policy banks, the China

Development Bank and the Export–Import Bank of China, now provide as much energy finance to foreign governments as do all the multilateral development banks combined.[49] Energy plays a key role in China's "Belt and Road Initiative" (BRI), a grand program to finance and build infrastructure in Asia, Africa, Europe, and beyond, which was announced in 2013. With projects in nearly 80 countries, the BRI aims to create a network of ports, railways, roads, pipelines, and industrial parks that will link China to cities from Bangkok to Rotterdam. No less bold is the ambition of State Grid, China's largest state-owned company, to create a global super grid called the "Global Energy Interconnection" (GEI) that will link every continent with undersea transmission cables to power the world with green electricity.[50]

The BRI and the GEI have strategic objectives. China hopes to reduce its dependence on energy and commodity imports that pass through chokepoints such as the Straits of Malacca and the South China Sea. China's infrastructure diplomacy could be as important to twenty-first-century geopolitics as the protection of sea lanes was to the hegemony of the United States in the twentieth century.[51] While the BRI has been embraced by many countries, it has also caused disquiet in some countries about China's growing influence.[52] Other concerns that have been expressed relate to indebtedness, transparency, the prominent role of Chinese contractors, and the environmental sustainability of these projects.[53] For now, China's energy loans are highly concentrated in fossil fuel extraction and power generation, especially coal.[54]

It should not come as a surprise that foreign investment in the energy sector is often politically charged. In oil- and gas-producing countries, there have been intermittent waves of "resource nationalism," with host governments frequently expropriating foreign firms.[55] In 2005, the US blocked the takeover of Unocal, an oil firm, by Chinese state-owned company CNOOC on national security grounds.[56] National security justifications were also invoked by authorities to prevent State Grid, the world's largest utility company and China's largest state-owned enterprise, from purchasing shares in certain electricity networks and utility companies, including in Australia (Ausgrid), Belgium (Eandis), and Germany (50Hertz).[57]

In some cases, cross-border energy infrastructure and cooperation plans are proposed as a way to stimulate integration and solve international conflicts.[58] The European Coal and Steel Community (ECSC) sought to harness the tools of war by bringing the coal and steel sectors under the authority of a regional organization. The *Ostpolitik* by Willy Brandt was a major factor that enabled the construction of East–West gas pipelines in spite of the Iron Curtain. In the 1990s, US officials promoted the idea of compelling Azerbaijan to build its major oil export pipeline through Armenia, as a means to bring peace to the South Caucasus.[59] Part of the motivation of Israel to import natural gas from Egypt was to increase the prospects for preserving peace.[60] The 1994 peace treaty between Jordan and Israel contained a proposal to construct grid connections between the two countries, plans that have nevertheless failed to materialize.[61] Geopolitical trust thus seems to be a precondition to building cross-border energy links rather than a consequence of it.

Resource wars and military conflict

Energy has been a key source of armed conflict, though probably in different ways than most people think. The dominant view is that oil and gas are scarce resources that have directly fueled most international conflicts and wars. The rapid depletion of conventional reserves coupled with the fast-growing energy hunger of countries like China and India is believed to trigger a "race for what's left." It could spark "resource wars" between major consumers.

In reality, there have been few actual "resource wars," that is, wars initiated primarily to territorially conquer oil and gas fields.[62] The first Gulf War may be an example, as well as Japan's invasion of Indonesia during World War II, but it is hard to find other examples beyond those. Oil may have been a factor in several other wars – the US invasion of Iraq in 2003, for instance, or the Iran–Iraq War – but even in those instances it never was the sole or even primary *casus belli*. Conquering oil fields or destroying enemy oil installations and supply lines might explain some of the military developments *in* major conflicts – like, for instance, the battle of Stalingrad in World War II – but that is fundamentally different from causing the conflict itself.

Much more common than full-fledged resource wars are so-called "oil spats": mild, usually non-lethal confrontations that state leaders quickly contain. They may consist exclusively of threats. On other occasions, countries may spar over oil by putting their armed forces on alert, mobilizing or moving troops to a contested border, or engaging in minor boundary violations. Contention between Libya and Tunisia in the mid-1970s centered on drilling platforms in contested waters. In 2000, Suriname's navy compelled a Guyanese oil rig to withdraw from disputed territory. This dynamic explains much of what is happening currently in the Mediterranean, the South China Sea, and the Arctic.[63]

Even if full-fledged oil wars are rare, great powers worry that adversaries could forcibly cut off oil trade at a critical moment – for instance, when they need to defend themselves in war. Great powers do not wait around to be victimized. Instead, they adopt anticipatory strategies to reduce vulnerability. In this vein, oil parallels nuclear weapons in their impact on international politics. Nuclear war has never occurred. Yet, the possibility of nuclear conflict was dangerous enough to profoundly shape US and Soviet conduct during the Cold War.[64]

Ever since Winston Churchill converted the Royal Navy from coal to oil in 1912, the safety of vital resource supplies has been a central feature of strategic planning for all major powers. The Eisenhower Doctrine of 1957 and the Carter Doctrine of 1980, though couched in the standard anti-Soviet rhetoric of the day, were principally intended to ensure continued US access to the Persian Gulf's prolific oil reserves. In 1983, President Reagan created the Central Command (Centcom) to guard Persian Gulf oil.[65] No other major power is capable of matching the United States when it comes to the global deployment of military power in the pursuit or protection of vital raw materials, but it is clear that China is very anxious about its high dependence on the Middle East and is preparing itself to better protect the sea lines of communication.[66]

One of the ways in which great powers reduce their vulnerability to oil coercion is by keeping oil in "friendly hands" and away from adversaries that may interfere with access. The strategy may include formal military alliances, establishing military bases, arms sales, and military aid. Such measures improve the petroleum producer's military capabilities and bolster the credibility of extended deterrence by establishing a commitment to the producer's defense.[67] This forms a major explanation for why the US developed a close relationship with Saudi Arabia at the end of World War II. In exchange for providing the US with so much of its oil, the Saudi government has relied on the US for protection against domestic revolt and external attack. In 1990, the US demonstrated the extremes to which it would go to defend Saudi Arabia when it deployed hundreds of thousands of its troops to ward off a possible attack by the Iraqi forces then occupying Kuwait.

The logical consequence of the resource war narrative, which guides how most people think about oil in international affairs, is that petrostates are likely to be the target of an attack rather than act as the aggressor. However, the evidence points in the opposite direction. Jeff Colgan's work has shown that "petrostates," where revenues from oil exports constitute at least 10 percent of GDP, have an "above average propensity to engage in militarized interstate disputes."[68] He found that "revolutionary petrostates" engaged in military conflict at a rate about 80 percent higher than non-petrostates over the period 1965–2001 – a phenomenon which he called "petro-aggression." His explanation was that revolutionary leaders are able to rely on oil export revenues to consolidate power and provoke international conflict. Thus the international trade in oil as currently structured places large amounts of money into a political system ill equipped to use it responsibly.[69] Examples of revolutionary petrostates are Iraq under Saddam Hussein (who invaded Iran in 1980 and Kuwait in 1990), Libya under Gaddafi (who engaged in four separate border wars with Chad, as well as a variety of militarized disputes with other countries such as Egypt, Tanzania, and the US), and Russia under Putin (who engaged militarily in Georgia, Ukraine, and Syria).[70]

Oil also impacts international security through its complicated links with terrorism. One of the world's most notorious international renegades, Osama bin Laden, cast the West's consumption of Persian Gulf energy as a central part of a complicated narrative that features the plundering of the Middle East's riches. Islamist insurgents appear to have taken these calls at least somewhat to heart, and have mounted episodic attacks against energy targets. In February 2006, for example, the Saudis thwarted an attack on the oil processing facility at Abqaiq.[71] In September 2019, a drone strike was more successful and hit Saudi oil facilities in Abqaiq and Khurais, taking away 5.7 million barrels or about 5 percent of daily production from global markets. It was the single largest daily oil supply disruption ever recorded. Terrorist and insurgent organizations like the Islamic State group, also known as ISIL/ISIS or Daesh, have also attempted to use oil as a revenue source.[72]

Colgan also developed a typology of "causal pathways" between oil and international conflict, depicted in table 3.1.[73] It lists no less than eight mechanisms linking oil to war, including classic resource wars, petro-aggression and terrorism.

Nuclear proliferation

Perhaps more than any other technology, nuclear power has been intrinsically linked to geopolitics for decades. Nuclear technology is

Table 3.1 Causal pathways between oil and international conflict

Dimension	Pathway	Causal mechanism	Example(s)
External and international: geopolitics and resources	Resource wars	Oil reserves raising the payoff of territorial conquest	Iraq–Kuwait, 1990; Chaco War; Japan, 1941
	Risk of market domination	Threat of conquest to ally or key territory	US–Iraq, 1991
Internal and domestic: politics in producing countries	Oil industry grievance	Presence of foreign workers creates grievances for state or non-state actors	Al-Qaida; Iran hostage crisis
	Petro-aggression	Oil reduces the accountability of leaders, lowering the risk of instigating wars	Iraq–Iran; Libya–Chad; Egypt
	Petro-insurgency	Oil income provides finances for actors to wage war	Iran–Hezbollah; Saudi Arabia–Afghanistan
	Externalization of civil wars	Oil creates conditions for civil war that then lead to foreign intervention or spillover	Libya–NATO; Angola–Cuba; Sudan–Chad
Internal and domestic: Access concerns in consuming countries	Transit route	Efforts to secure transit routes create a security dilemma	Sudan; South China Sea; Strait of Hormuz
	Obstacle to multilateralism	Importers attempt to curry favor with petrostates to prevent multilateral cooperation	US–China friction over Iran; Sudan

Source: Modified from Colgan, J. D. (2013) Fueling the fire: Pathways from oil to war. *International Security*, 38(2), 147–80.

dual use in nature, meaning that it can be used to produce nuclear power or to build nuclear weapons. The same nuclear plant that produces electricity also produces plutonium, which could be used as a weapons fuel. If a country develops uranium enrichment technology to produce low-enriched uranium fuel for their reactors, they also have the capability to produce highly enriched uranium for use in nuclear weapons.[74]

The origins of nuclear energy technology also lay in war. Under the code name "Manhattan Project," the US was the first country to develop atomic bombs, which were used in Hiroshima and Nagasaki in 1945.[75] After the war, in the emerging global context of the Cold War, the development of nuclear technology continued, primarily for weapons and other military applications such as nuclear-powered submarines.

By 1953, three countries had acquired nuclear weapons: the US, the Soviet Union, and the UK. It led US President Eisenhower to deliver his famous "Atoms for Peace" speech at the United Nations, spelling out the necessity to repurpose nuclear weapons technology to peaceful ends. The US offered nuclear technology and fuels to developing countries for civilian purposes, which meant, primarily, the generation of electricity with nuclear power.[76] One of the motivations for this speech was that Eisenhower had come to realize that it would be impossible to stop the spread of the technology.

The central theme behind the Atoms for Peace project was to show that the power of the atom could be converted from a terrifying military force into a benign commodity. The role of the government was to be a custodian of atoms.[77] Eisenhower's Atoms for Peace program earmarked $475 million in funds to promote nuclear power abroad where it was believed the United States had a moral duty to lead a global transformation in which atomic energy would morph deserts into jungles, produce an infinite supply of freshwater, allow production of cheap fertilizer that would eliminate famine, enable humans to live comfortably under the sea and in outer space,[78] create new lands literally flowing with "milk and honey,"[79] and convert Africa "into another Europe."[80] In an effort to make this a reality, by the end of 1957 the United States had signed bilateral Atoms for Peace agreements with 49 countries and US firms had sold 23 research reactors overseas.[81]

Eisenhower's speech paved the way for the commercialization of nuclear technology, while providing political cover for a large-scale nuclear weapons build-up at home. It also laid the foundations for the creation of the International Atomic Energy Agency (IAEA) in 1957 and the Non-Proliferation Treaty (NPT) in 1968. The implicit

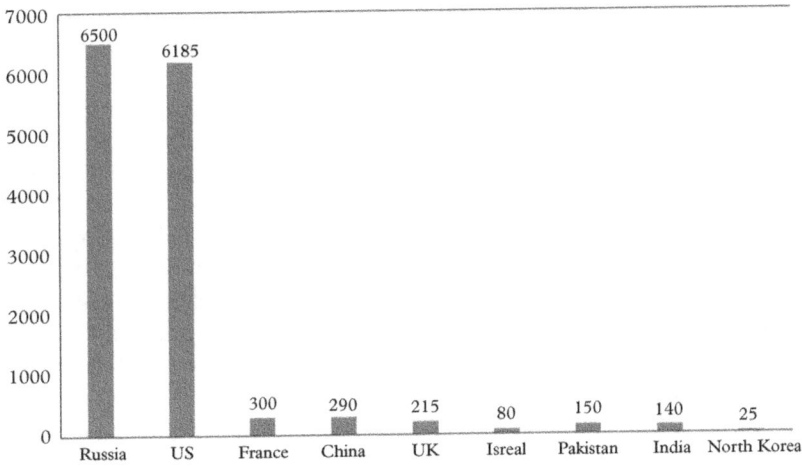

Figure 3.2   Estimated global nuclear weapons arsenal, 2019
Source: Adapted from Federation of American Scientists, *Status of World Nuclear Forces.*

bargain of the NPT was that, in exchange for access to peaceful nuclear assistance from the countries holding nuclear technology, non-nuclear weapon states would not pursue nuclear weapons. By the time the NPT was signed, the countries that possessed nuclear weapons had grown to five, with the addition of France and China in the 1960s.

In spite of this bargain, the five authorized nuclear weapons states under the NPT have not seriously pursued disarmament.[82] Despite the Cold War being over for three decades, close to 14,000 warheads form the combined stockpile of nuclear weapons. Figure 3.2 shows that the United States and Russia still have more than 6,000 weapons each.

Meanwhile, states not listed among the five authorized under the NPT to possess nuclear weapons have not paid a significant price for their pursuit of weapons capabilities. States that were never party to the NPT – Israel, India, and Pakistan – have at various stages after 1968 attained nuclear power status, thus breaking the nuclear monopoly of the five nuclear weapon states recognized in the NPT. Among parties to the treaty, North Korea withdrew from the NPT in 2003, and has embarked on its own ambitious nuclear program, spurring some in neighboring Japan to advocate going nuclear as deterrence against a threatening North Korea. Meanwhile, Iran, another signatory, has been exercising its right under the NPT to develop "peaceful" nuclear technologies. Attempts by Libya and Iran to acquire nuclear weapons have been stymied by international sanctions.

The NPT has been significantly weakened by a number of bilateral deals made by NPT signatories, notably the United States. In October 2008, for example, the United States signed a bilateral nuclear deal, the "123 Agreement," with India. The deal allows India to receive nuclear technologies and fuel from the Nuclear Suppliers Group, an international consortium that controls trading in nuclear weaponry, fuel and technologies; it also exempts India from having to sign the NPT.[83] These deals contribute to the notion that a "double standard" exists among NPT signatories where they will flout the norms of the treaty whenever it suits them.[84]

The threat of states or terrorist organizations actually using nuclear weapons and materials in an attack still represents one of the most existential risks that humanity faces. There is no shortage of terrorist groups eager to acquire the nuclear waste or fissile material needed to make a crude nuclear device or a dirty bomb.[85] The risks are not confined to the reactor site. All stages of the nuclear fuel cycle are vulnerable, including attacking a nuclear power reactor directly; assaulting spent fuel storage facilities; and intercepting nuclear materials in transit.

Geopolitics of the energy transition

A global transition from fossil fuels to renewables would surely have major geopolitical ramifications. Fossil fuel reserves are concentrated in specific geographic locations, while renewables are much more dispersed – they are available in one form or another in most countries. Renewables take the form of flows (of wind, water, sun) rather than stocks, which means that they cannot be exhausted and are harder to disrupt. In the words of former US President Jimmy Carter: "no one can ever embargo the sun, or interrupt its delivery to us."

A global shift away from fossil fuels and toward renewables and other low-carbon technologies is likely to lead to global power shifts. The obvious losers are those countries highly dependent on fossil fuel export revenues – think of countries like Angola, Kuwait, Nigeria, or Venezuela. These countries will face economic, social, and political risks if they do not take steps to transform and diversify their economies. The winners are the importers of fossil fuels.

If they switch from imports to producing their energy domestically, they will improve their trade balance and enjoy more foreign policy autonomy.

Three other types of countries stand to gain: mineral-rich countries like Bolivia or Mongolia; countries with a high technical potential for renewables like Chile and Australia (which could become significant exporters of renewable fuels such as hydrogen or ammonia made from solar and wind); and, especially, countries that are at the forefront of the clean energy technology race. Currently, China is leading the pack by some distance. It is the world's largest producer, exporter, and installer of solar panels, wind turbines, batteries, and electric vehicles. It is also the world's largest investor in clean technologies and it leads the world in renewable energy patents. Yet, given the decentralized nature of renewables, the energy transformation could empower new actors such as citizens, cities, and regions who will have the ability to generate their own energy. Altogether, this adds up to a dispersion of political power.

The energy transition will create new geographies of trade and new relations between states. In general terms, we will see a shift in energy trade from global fuel markets to regional grids. Some maritime trade routes – and so-called chokepoints – will become relatively less important, while alliances built on fossil fuels are also likely to change. Electricity, biofuels, and other materials critical for renewable technologies will gain in significance but they are unlikely to pose the same geopolitical risks or even gain the strategic importance of oil and gas. The energy transition, while reducing competition for oil and gas, does raise new security concerns with regard to cybersecurity and access to minerals. For instance, most of the world's cobalt, which is used in batteries, comes from the Democratic Republic of the Congo (DRC), a country that is suffering from poor governance and conflict.

The transformation is also likely to affect the root causes of geopolitical conflict. It may create social dislocation, for instance when a lot of coal miners lose their jobs. The yellow vest movement is another example of new social divisions being created. There is the risk of a global financial meltdown, caused by the risk of stranded assets. On the upside, renewables have the potential to mitigate climate change, which is widely regarded as a global security threat. Finally, there is the possibility of developing countries leapfrogging. Just as a number of countries have skipped landlines and moved to mobile phones directly, off-grid renewables do not require the big infrastructure investments of the past.[86]

Conclusion

This chapter sought to uncover the intricate ways in which energy systems influence foreign policy, military conflict, and broader geopolitical relations. It should be clear that energy has left a large imprint on the security of nations, their alliances, and their overall stance in the international system. While energy has been used as a weapon in foreign policy, notably in the first Arab oil embargo, it should be noted that the international markets for oil and gas today are far more robust and integrated than they were a couple of decades ago.[87] Yet, this does not render political blackmail, extortion, and interference less likely. As long as energy is central to our modern lifestyles, energy systems will always be a key feature of the geopolitical landscape. As we will show in the next chapter, energy also affects the fortunes of nations.

4

Energy and the economy: Powering growth and prosperity?

The history of modern economic development is largely a story of humankind harnessing ever-greater amounts of energy. From a time when people relied only on their own muscle power, or those of slaves, we have gradually learned to master fire, domesticate animals, and make use of wind and water power. A turning point came around 1800, when the fossil fuel revolution took off, unleashing unprecedented amounts of energy. The average person now consumes about 50 times more energy than the hunter-gatherer did some 100,000 years ago – and we put in much less effort.[1] Fossil fuels can be regarded as modern "slaves" that do all the work for us. One barrel of crude oil contains the energy equivalent of two slaves working for you around the clock for a full year.[2]

The surge in energy use that began around 1800 has enabled what economist Deirdre McCloskey has called "the Great Enrichment," the extraordinary growth in wealth enjoyed by an ever-larger part of the world population over the last 200 years.[3] The benefits to humankind have been immense, extending life expectancy, increasing food output to sustain burgeoning populations, and lifting the standard of living for many people to levels not even kings could aspire to just a few centuries ago.[4] Fossil fuels have made a whole range of luxury goods available to ordinary consumers, including air travel, central heating and air conditioning, supermarkets with a huge choice of products, and the internet.[5]

As a result of this evolution, today, many of us live in high-energy societies, powered mainly by fossil fuels. Oil, coal and gas are still the backbone of the world economy and modern life styles, and they are an underappreciated factor in the process of globalization.[6] Clearly, energy is not just "another sector" in the economy like transport, agriculture, or manufacturing. Energy is one of the fundamental factors of production, alongside labor, capital, and technology.[7] It is what classical economists used to call a "basic good": directly or indirectly, it enters the production of *every other* produced good or service.[8] It provides the fuel that drives our transportation system, heats our homes and offices, and powers our factories. The bad news is that our "energy slaves," as they work for us, also heat the planet, pollute the air that we breathe, and impact lives and livelihoods of people, mostly in places far away from the consumer.

Fully cognizant of these interlinkages, this chapter looks at the economics of the global energy system. It proceeds in three parts, each focusing on a different type of actor: energy-importing countries, energy corporations, and energy-exporting countries. For importers – the large majority of countries in the world – the critical question is how energy prices affect the world economy. For corporations, we explore the different roles played by state-owned and private companies across various segments of the value chain of the energy business. For exporters, this chapter explores the evidence, explanations, and remedies for the so-called "resource curse."

Energy and the world economy

A unique sector

The energy industry differs from the rest of the economy in at least three important ways.[9] First, energy is the world's most capital-intensive sector. According to Deloitte, between 2009 and 2013, energy and resources accounted for almost 50 percent of total new capital raised by all nonfinancial industries.[10] It is not exceptional for oil and gas projects to cost billions of dollars. The development of the offshore Kashagan oil field in Kazakhstan's zone of the Caspian Sea, for example, is estimated to cost a staggering $116 billion – more than five times the cost of building the Channel Tunnel that links the UK to France.[11] The oil and gas sectors have traditionally attracted the largest chunk of energy investment, but the power sector has overtaken it since 2016.[12]

Due to capital intensity and scale of energy projects, energy investments typically cast a very long shadow. The average lifespan of energy-related capital stock ranges from around two decades (for wind turbines), 40 years (for pipelines and nuclear power plants), 50 years (for coal-fired power plants) to over 80 years (for large hydropower plants) – see figure 4.1. For large oil fields, it takes on

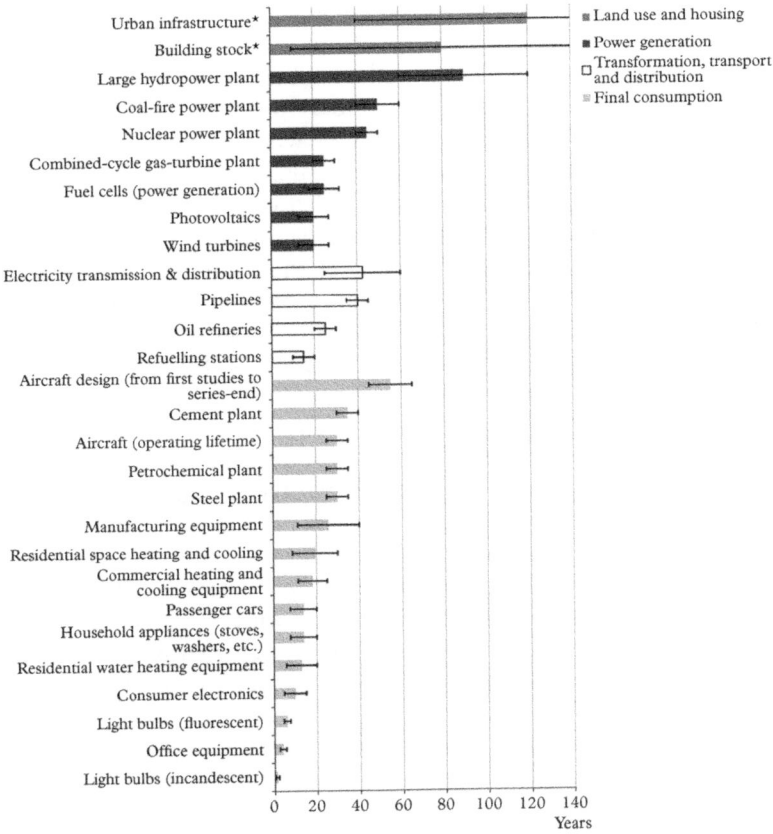

Figure 4.1 Average technical lifespan of energy-related capital stock

Source: IEA (2008) *World Energy Outlook 2008.* Paris: OECD/IEA, p. 75.

average some 5 years from first discovery to production,[13] and up to 10 years for technically demanding reservoirs such as offshore fields.[14] Moreover, once production has started, there is usually an incentive to keep producing, even when revenue is insufficient to do more than recover operating costs.[15] As a result, there is a high degree of inertia built into the energy system.

Second, few other sectors of the global economy are as internationalized as the energy business. Energy resources such as crude oil and natural gas are very concentrated geographically. Unlike factories, oil fields are where they are and cannot be "created" elsewhere. The production structure of oil and gas thus fundamentally differs from that of manufacturing and service industries.[16] It means that energy resources are often traded across borders before they are consumed. Around 90 percent of all the uranium that is mined, 70 percent of all the oil that is pumped out of the ground, 30 percent of all the natural gas that is extracted, and 18 percent of all the coal that is dug up are exported from the country where they are produced.[17] Consequently, energy takes up a large share of international trade. Comprising almost 20 percent of world trade, fossil fuels represent the largest product category in trade value terms.[18]

If we move away from just looking at the energy resources themselves and, instead, adopt a systems perspective, it becomes clear that many countries are dependent on foreign nations not just for the supply of various fuels, but also for a variety of technologies, services, capital, and access to markets.[19] In this regard, no country can truly claim to be "energy independent." Even energy-rich countries are often entangled in seemingly paradoxical dependencies. Iran sits on the world's fourth largest oil reserves but has struggled for years to meet its domestic fuel needs due to a lack of refining capacity, which forced it to import large volumes of gasoline.[20] Venezuela, in spite of having the world's largest oil reserves, imports small volumes of oil in the form of naphtha from the US to mix with its extra heavy crude oil from the Orinoco belt to make it exportable – that is, at least, until US sanctions imposed against the Maduro regime abruptly cut off supply in January 2019.[21]

This international interdependence is compounded by another feature, which is that energy fuels and electricity require specific modes of transport. Factors such as labor, capital, technology, and even raw materials can be transported using general-purpose infrastructures, such as roads, satellite communication, and bulk carrier ships. Many forms of energy, by contrast, require dedicated infrastructure and transport modes, such as pipelines, LNG ports, or power grids, to reach their market. No other industry features such an internationalized "midstream" sector as energy.[22] Many of these energy networks, especially natural gas and electricity transmission and distribution systems, exhibit natural monopoly features. Due to the high sunk costs in large-scale energy transport infrastructure, there are high barriers to entry for new competitors, and it can be less efficient to have multiple providers. Not so long ago, the EU and Russia supported the construction of two rival gas pipelines in Southeast Europe – Nabucco and South Stream. Both projects have now been shelved but it was always clear that constructing both pipelines would not have been economically efficient.[23]

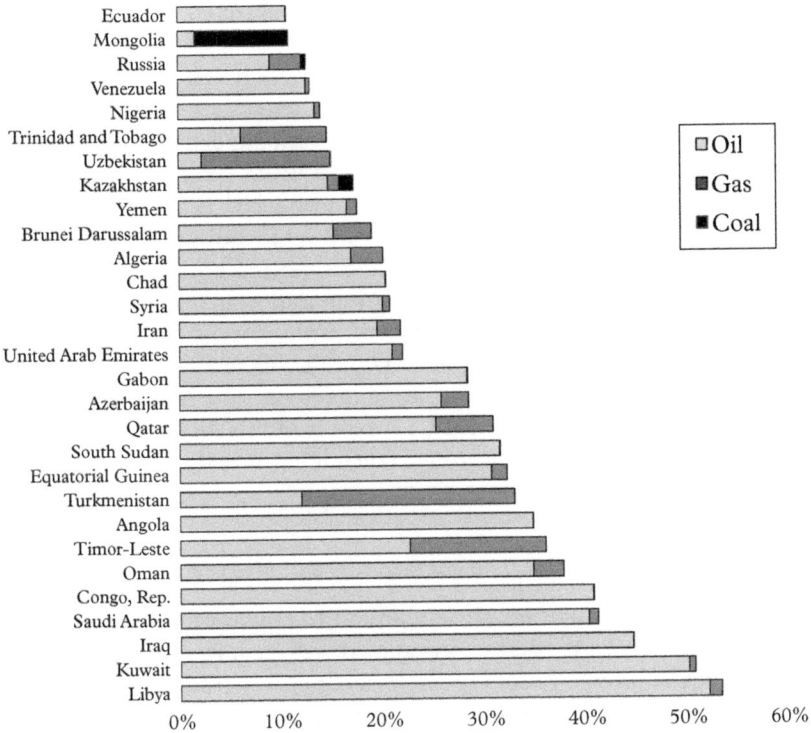

Figure 4.2 Fossil fuel rents as a percentage of GDP
(average 2007–2016)

Source: Adapted from World Bank, World Development Indicators.

A third distinctive feature of energy markets and investment is the importance of economic rent and of conflict over rent distribution. Resource rents are typically defined as the difference between revenues and extraction cost. Rent exists in all markets but, as Albert Bressand argues, "rent is far more pervasive, lasting, and protected in energy markets than in most other markets."[24] The World Bank estimates resource rents at about US$4 trillion annually, or 7 percent of global GDP.[25] Almost 30 countries are highly dependent on fossil fuel rents. As figure 4.2 shows, oil rents are many orders of magnitude larger than gas and coal rents. These rents are often key for the state and political survival of the ruling elite. This is especially true of so-called "rentier states," whose political institutions are built around the capture and distribution of resource rents by state agencies.[26]

Rents are also important for importing countries. Most countries in the world do not – and cannot – have domestic production of oil or gas as they lack the resources. Yet, they often put in place consumption taxes which, for countries with zero extraction, are identical to import tariffs.[27] It is often overlooked that OECD governments earn more from oil taxes than OPEC countries earn from taxing their foreign oil sales. Over the period 2013–17, for example, OECD countries earned an average of $2,445 billion per year from retail oil taxes, whereas OPEC members earned an average of $722 billion per year.[28] This revenue stream is important for governments of consumer countries. In 2016, for example, energy taxes amounted to almost 5 percent of total tax revenue collected in the European Union, with most of it coming from taxes on oil products.[29] These petroleum taxes levied by OECD governments have long been a thorn in the side of oil exporters, who denounce them as an unfair tool to snatch rent away from them.

Industrialization and energy demand

Patterns of energy use are closely linked to stages of development at the national level. As their economies develop, states tend to ascend an "energy ladder," reducing their reliance on traditional biomass while electricity use and per capita energy use increase. At the same time, the energy intensity of economic growth declines as countries grow richer.

For instance, developing countries such as India or China need three times as much energy as the developed nations to produce one unit of GDP.[30] Developing countries are energy intensive, partly because of various inefficiencies, such as the widespread subsidization of energy. But the main explanation lies in the nature of economic growth.[31] Developing countries are rapidly building their infrastructure, including roads, houses, schools, factories, sewage, and power systems. All of this requires a lot of energy and energy-intensive materials such as cement and steel.[32] The developed nations, in contrast, have more mature economies that are less dependent on heavy industry. They tend to have more technically sophisticated light industry and a large service sector.

To some extent, developed countries such as the UK and Denmark have been "outsourcing" or "offshoring" some of their energy consumption (and related emissions) to developing countries, as goods once manufactured domestically are now imported from countries like China. Still, CO_2 emissions from the burning of

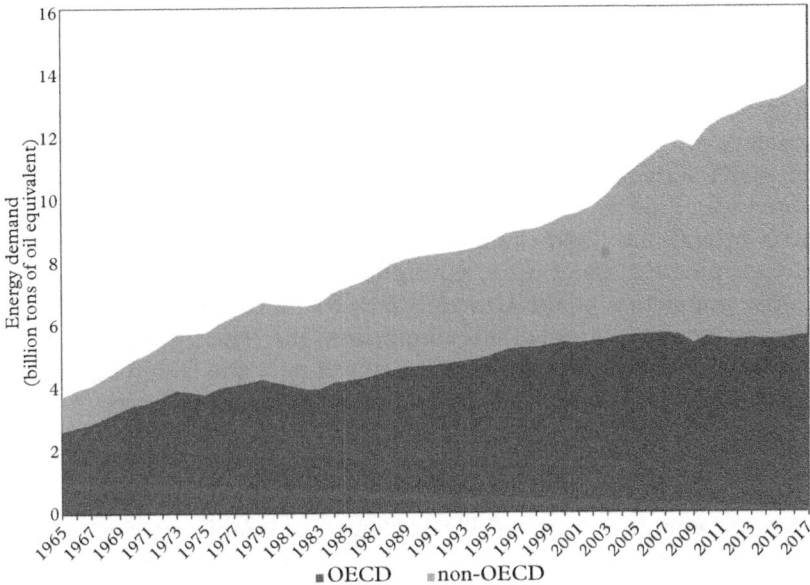

Figure 4.3   The shifting geography of global energy demand
Source: BP (2019) *Statistical Review of World Energy.*

fossil fuels are conventionally attributed to the country where they are produced, which is only a single point in a much broader value chain. Consumption-based accounting of greenhouse gas emissions would lead to a somewhat different picture, as up to a quarter of global emissions stem from the production of internationally traded goods.[33]

The shift of energy demand from advanced to emerging economies is one of the biggest ongoing transformations in global energy markets. Since the start of this century, energy consumption in developing countries has grown strongly, while in OECD countries it peaked in 2007 (see figure 4.3). This trend is set to continue over the next few decades. According to the IEA's central scenario, global primary energy demand expands by over 25 percent between 2017 and 2040 – and it could be twice as large without improvements in energy efficiency.[34] India is the single largest source of global energy demand growth in this period, followed by China and then Africa. Energy demand in the US is projected to stay flat, while it falls in Japan and the European Union. How to meet this anticipated growth in global energy demand at acceptable cost and while also reducing carbon emissions is one of the biggest challenges of the twenty-first century.

The stellar economic growth in countries like China and India has caused unprecedented structural transformation. Hundreds of millions of people have switched from agriculture, a low-energy-intensive activity, to construction and industry, high-energy-intensive activities. This process of industrialization has gone hand in hand with urbanization and changes in lifestyles. China has a burgeoning urban middle class of up to 400 million consumers, larger than the entire population of the United States.[35] China seems to be following the familiar pattern and is climbing up the "energy ladder": since recently, the country is growing at a slower pace and is shifting investment toward services and higher value-added manufacturing.[36]

Energy prices and subsidies

Because energy is required for any given economic activity, energy prices are key economic parameters. They influence consumer spending, industrial cost structures, and ultimately global inflation and economic growth. The oil price is especially important for the world economy. With an annual production value of more than $3,500 billion (corresponding to 4.8 percent of global GDP) and an export value of $2,700 billion (equal to some 12 percent of global goods trade), oil dwarfs all other leading primary commodities.[37] Oil has also special significance because of its indispensability – it is hard to replace it in the short and medium term as a fuel for transport on roads, in the air, and across the seas.

Oil prices can have a significant effect on the macroeconomy. Nine out of ten of the US recessions since World War II were preceded by an upward spike in oil prices,[38] although there continues to be a heated debate about the extent to which oil shocks contributed to these recessions. Conversely, a drop in oil prices might give a shot in the arm to the world economy. Yet, the effect varies significantly for each national economy. Dearer oil imports can impose a huge burden on the trade balance of energy-importing countries. In 2017, the global cost of gross fossil fuel imports was over US$2 trillion (over 10 percent of global merchandise import costs).[39] For countries as diverse as India, Tanzania, Bahrain, Belarus, and Greece, fossil fuel imports represent more than 30 percent of all merchandise imports (see figure 4.4).

From the perspective of energy-importing countries, high energy import bills simply transfer large amounts of wealth abroad.[40] High

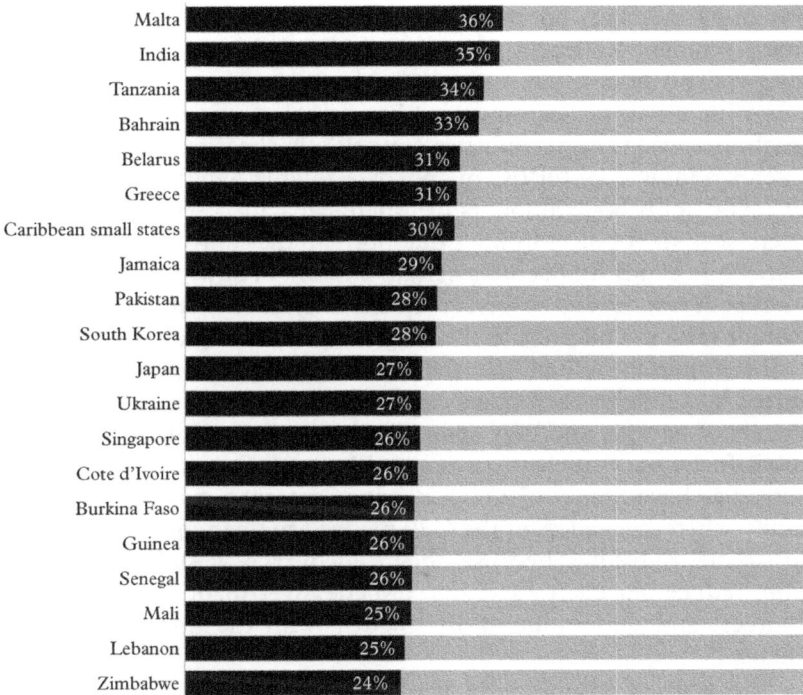

Figure 4.4 Share of fossil fuel imports in all merchandise imports (average 2012–2017)

Source: World Bank, World Development Indicators.

dependence on imported fuels makes them also vulnerable to price swings, often as a result of remote geopolitical events. High oil prices can stifle economic growth by importing inflation, curbing consumer spending, and raising the cost of doing business. Lower oil prices, conversely, can be a real shot in the arm of the economy of importers. For example, the large drop in oil prices in 2014 has led to a one-off boost to the EU economy and an estimated GDP increase of 0.8 percent in 2015 and 0.5 percent in 2016.[41] The impact of volatile fossil fuel prices on the economies of exporting countries is discussed further below, as part of our discussion of the resource curse.

In Western public opinion, it is often assumed that OPEC can raise oil prices by agreeing to production cuts. The US Congress has therefore threatened repeatedly to adopt the so-called No Oil Producing and Exporting Cartels (NOPEC) Act, which would allow the US attorney general to sue the oil producers group or any of its members on grounds of violation of anti-trust law.[42] Yet, many scholars are highly skeptical of OPEC's ability to manipulate long-run oil prices. Jeff Colgan, for instance, claims that the view of OPEC's output decisions as a key driver of prices is a persistent myth, because OPEC countries consistently cheat on their quotas.[43] Even so, individual OPEC members – notably Saudi Arabia (which typically holds large amounts of "spare capacity") – can unilaterally exercise market power in ways that prop up the short-run price of oil. Yet, while Saudi Arabia has a certain amount of power to stabilize the market in response to *temporary* shocks to either demand or supply, it cannot raise the oil price in the long term by cutting output.[44]

It is important to highlight the fact that energy prices are often distorted by subsidies. Subsidies can be given to the consumers or producers of energy, and they span all areas of the energy sector, from fossil fuels, over nuclear, to renewables and electricity. Fossil fuel subsidies have come to the forefront of international debates since 2009, when the leaders of the G20 pledged to phase out "inefficient" fossil fuel subsidies because they "encourage wasteful consumption, reduce our energy security, impede investment in clean energy sources, and undermine efforts to deal with the threat of climate change."[45]

The scale of subsidies for fossil fuels is massive. The IEA estimates that fossil fuel consumer subsidies alone amounted to US$548 billion annually in 2013 (figure 4.5). This is equivalent to four times the level of official development assistance aid from OECD countries in 2013 (US$134 billion) or more than four times the level of total financial support to renewable energy (US$121 billion).[46] As figure 4.5 reveals, fossil fuel subsidies tend to follow the trajectory of oil prices – including the rebound in 2017. Still, the 15 percent rise in subsidies in 2017 was considerably less than the 25 percent rise in oil price.[47]

There are also "implicit" subsidies to fossil fuels, because their price does not reflect the environmental and social costs they generate in the form of climate change, local air pollution, and even traffic congestion and accidents. Taking all of these externalities into account, the International Monetary Fund (IMF) estimates global fossil fuel subsidies to actually be in the order of $5.2 trillion, or the equivalent of 6.5 percent of global GDP.[48] This is much higher than the IEA's figure of $300 billion (in 2017).

There can be good reasons for governments to make energy more affordable, particularly for the poorest and most vulnerable groups. But often, these subsidies are poorly targeted, disproportionally

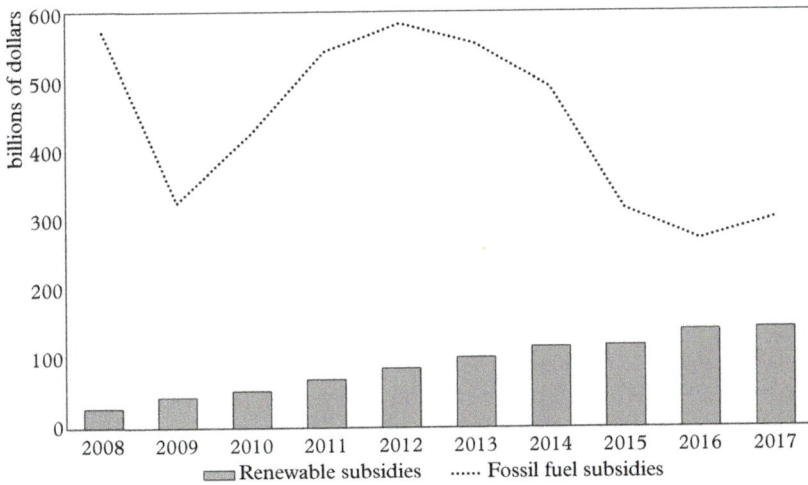

Figure 4.5   Comparing global subsidies for fossil fuels and renewables

Notes: Fossil fuel subsidies include just consumption subsidies, not production subsidies. Renewable subsidies comprise the cost of support mechanisms on a global basis for renewables-based electricity, excluding subsidies to hydroelectric generation.

Source: IEA, Energy subsidies database: https://www.iea.org/weo/energysubsidies/.

benefiting wealthier segments of the population that use much more of the subsidized fuel. Such subsidy policies encourage wasteful consumption, push up emissions, and strain government budgets.[49] In Indonesia, for instance, fossil fuel subsidies accounted for no less than 28 percent of the government's expenditure in 2014, one year before it abolished gasoline subsidies and put a cap on diesel subsidies.[50]

Fiscal strains, often a result of high international oil prices, can galvanize states into action. Between 2015 and 2018, no less than 50 countries have undertaken efforts to reform fossil fuel subsidies. Yet, phasing out energy subsidies often runs into trouble.[51] In some countries, consumption subsidies are part of the "social contract" between governments and the population.[52] In other cases, reform measures may get bogged down over popular protests, opposition from interest groups,[53] or the credibility of the institutions entrusted with implementing reforms.[54] Reform efforts have led to unrest in at least 19 countries in the decade since 2006.[55]

As far as renewable energy subsidies are concerned, in many markets, there has been a shift from subsidies to auctions and other types of support mechanisms. In 2017, more than 20 percent of new solar projects that received support were selected on the basis of competition, together with about 30 percent of onshore wind and 50 percent of offshore wind projects.[56]

Nuclear power plants are also heavily subsidized. Nearly all new builds in recent years have taken place in markets where electricity prices are regulated or where government-owned entities build, own, and operate plants. In competitive markets, the risks in constructing new nuclear plants have been too large to attract investment, though some governments have offered subsidies to mitigate these risks.[57] A case in point is the UK government's support for the expansion of Hinkley Point C nuclear power plant. In 2013, the price UK electricity consumers would pay for Hinkley C's power was revealed: £92.50 per MWh – more than double the wholesale power price at that time, adjusted for inflation and guaranteed for 35 years.

The energy industry

As sketched in chapter 2, Western private oil corporations have played a key role in the historical development of the oil market. While these international oil companies (IOCs) are still very important actors today, national oil companies (NOCs) have come to dominate many aspects of the global energy scene. Here we discuss each in turn.

Private companies

Given the energy sector's heavy capital intensity, it should come as no surprise that private energy corporations are among the world's largest companies by revenue – in 2017, a year of relatively low oil prices, six of the top ten largest publicly-listed companies in the world were energy companies. They included State Grid, Sinopec Group, China National Petroleum Corporation, Royal Dutch Shell, BP, and ExxonMobil.[58] At the end of 2018, ExxonMobil had a market value of $289 billion. If it were a country, ExxonMobil would be the 42nd biggest economy in the world.[59]

Oil behemoths like ExxonMobil are notorious for their lobbying power. In his book *Private Empire*, journalist Steve Coll accounts Exxon's travails from the 1989 Exxon Valdez oil spill to 2010, exposing how the corporation is truly a state within the state.[60] But it would be wrong to talk about "the oil industry" as though this is a homogeneous bloc. Alongside integrated oil and gas companies like Exxon, Shell, or BP, there is now also a large oil and gas service industry, consisting of companies like Schlumberger and Halliburton that provide oil and gas field development services to production companies selling oil to end users. Also, the interests of the refining sector and the upstream oil companies are not always aligned.[61] This was evident in the US in 2015 when refineries lobbied against the lifting of the US crude oil export ban because they benefited from low crude prices in the US, whereas the oil companies wanted to see the ban lifted.[62]

Private oil companies face a number of challenges. First, they need to replenish their reserves constantly but they cannot access many low-cost oil fields as these are mostly in the hands of national oil companies. So, they have to veer off to other fields that might be more difficult to develop, either technically (e.g., Arctic or deepwater fields), politically (e.g., the conflict-prone Niger Delta), or environmentally (e.g., shale or tar sands). The pressure on these companies to keep up their reserve levels was brought painfully to the fore in 2004, when it was revealed that Shell had over-reported its proven oil reserves by 4.5 billion barrels, around 23 percent of its total reserves, slashing its market value and leading the US authorities to impose a $120 million fine.[63]

Second, these companies face growing public scrutiny over some of their corporate wrongdoing, such as when they deal with authoritarian regimes or are deemed responsible for environmental disasters or human rights abuses. Shell, which has major oil and gas operations in Brunei, faced global criticism and a global business backlash in April 2019, after the tiny south-east Asian nation introduced harsh anti-gay laws.[64] Sometimes it does not just stay with public condemnation. In 1995, Shell faced a consumer boycott after Greenpeace had launched a campaign against the company's plan to sink an old oil rig in the North Sea.[65] BP, the company responsible for the 2010 Deepwater Horizon oil spill, had to pay $65 billion in court-ordered compensation payments.[66] Some of the biggest oil and gas companies are now being sued by US states, cities, and even children, seeking compensation for climate change damages.[67] This string of litigation follows revelations that major oil companies like Exxon knew of the dangers of climate change for decades, but still spent millions on misinforming the public as well as regulators – a practice brilliantly exposed in the book *Merchants of Doubt*.[68] From a legal perspective, fossil fuel companies that fail to be transparent about the damage their operations pose to the world's climate could be liable in the same way as tobacco companies were for not telling the truth about the health damages from smoking.[69]

Another challenge these oil firms face is increasing pressure from shareholders and investors for greater disclosure about how their business plans align with the Paris climate goals. While climate risk disclosures are steadily becoming routine in the industry, Shell went further in 2017 by setting targets to reduce the emissions from the energy products it sells. In 2018, it decided to link executive pay to meeting those targets.[70] In February 2019, mining giant Glencore, the world's top coal exporter, yielded to investor pressure to cap its coal production.[71] All of these trends indicate major shifts in the fossil fuel industry's social license to operate.

If oil companies face existential risks from the energy transition, electric utilities might appear as the big winners, especially if the future is increasingly electric. The reality is more nuanced, however. Several oil companies are taking serious steps to diversify their activities into low-carbon technologies, renewables, and electricity.[72] Danish Oil and Natural Gas (DONG) changed its name to Ørsted after selling off its oil and gas business. Total is diversifying into solar and battery technology. Shell, for example, aims to become the world's largest electricity company by the 2030s.[73] BP and other oil companies are buying into electric vehicle charging infrastructure. In general, the European majors have been more proactive and the American companies more reactive on the issue of climate change and the rise of renewables.[74] It remains to be seen to what extent these moves mark the advent of a new era for Big Oil. The example of BP offers a cautionary tale: in 2000, it branded itself as "beyond petroleum," but a couple of years later it wound down its solar, bio-fuels, and wind activities.[75]

The electric utilities, meanwhile, are under similar pressures to adapt as they struggle with zero-marginal cost and decentralized renewables on the grid. German utilities RWE and E.ON have more than halved in value in recent years and France's EDF has seen its market value fall from over €100 billion to less than €20 billion.[76] As the conventional electric utility model is gradually becoming obsolete, new companies will probably enter the market. A key example is Tesla, which is not just building cars but also batteries in its so-called "gigafactories." In 2017, barely seven years after its initial public offering, Tesla became the most valuable US car maker, over-taking General Motors, a company over 100 years old. Some global technology companies like Apple and Microsoft already source 100 percent of their electricity from renewable energy,[77] while many more global companies have committed to the same goal, including IKEA, Walmart, and Tata Motors.[78] These trends indicate that the impending energy transition could very well lead to a corporate shakeout.

National oil companies

In terms of oil reserves, private oil companies are dwarfed by the NOCs from oil-producing states, who play critical roles in their home-country economies as well as in global oil and gas markets. Today, NOCs command approximately 90 percent of the world's oil reserves and 75 percent of production, as well as many of the major oil and gas infrastructure systems.[79] This situation is a complete reversal from the pre-nationalization era, when the Seven Sisters controlled about as much of oil reserves. Figure 4.6 lists some of the major NOCs by oil and gas production (expressed in terms of billions of oil equivalent), for which there was available data. The list is topped by Saudi Aramco (Saudi Arabia) which in 2019, for the first time, lifted the curtain on its financials, revealing its net income in 2018 to be over $100 billion, making it by far the world's most profitable company.[80]

As figure 4.6 also reveals, a new category of NOCs has gained importance, namely the national oil companies of net oil-importing countries. The most prominent examples are CNPC, CNOOC and Sinopec (from China) and ONGC (from India). Beginning in the late 1990s, Asian and Russian NOCs have begun to invest hundreds of billions of dollars in foreign oil and gas assets. This "buying spree" was met with great suspicion in the West because these NOCs may benefit from significant diplomatic and financial support from their home governments. Moreover, these NOCs often invest in countries deemed hostile to Western interests and values. In 2019, for instance, the Russian giant Rosneft is allegedly providing a financial lifeline to help Nicolás Maduro cling on to power in Venezuela.[81] For some time, there has been concern that Chinese NOCs, by making equity investments, were taking oil out of the market, but these concerns were ungrounded.[82]

While there is much difference among the NOCs, research has found that they do not always have the incentive to develop their reserves at the same pace as investor-owned IOCs. NOCs also pursue goals that are not necessarily market-oriented but help to pursue the objectives of their host governments. The goals of NOCs often include employing citizens, furthering a government's domestic or foreign policies, generating long-term revenue to pay for

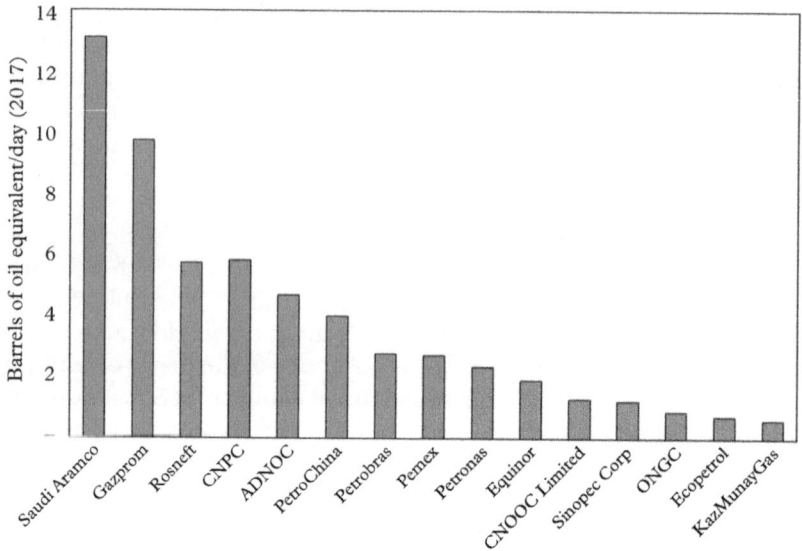

Figure 4.6   Top 15 national oil companies by oil and gas production
Source: Natural Resource Governance Institute, National Oil Company
Database, 2019.

government programs, and supplying inexpensive domestic energy
at subsidized prices.[83]

Oil-exporting countries and the resource curse

At first glance, discovering a giant oil or gas field seems like a
momentous stroke of good fortune for a country, akin to winning
the "geological lottery." Unfortunately, like real lottery winners,
countries often find themselves struggling to manage enormous sums
of money from newfound wealth. Meanwhile East Asian states with
few resources have enjoyed rapid economic growth in the second
half of the twentieth century. This observation has given way to the
thesis that resource booms might be more of a curse than a blessing.
The concept of the "resource curse" was coined by Richard Auty
in the early 1990s,[84] and has since given rise to a voluminous body
of work.[85] Put simply, the resource curse thesis suggests that countries with large reserves of natural resources often perform worse in
terms of long-term economic growth than comparable resource-poor
countries.[86]

It is important to understand what the resource curse thesis is, and what it is not. First, the resource curse is by no means an exclusively economic phenomenon. Researchers have also tied resource wealth to human development indicators such as poverty and undernourishment,[87] schooling,[88] gender inequality,[89] corruption,[90] authoritarianism,[91] and an increased frequency of conflict and war.[92] Second, the resource curse has been associated not just with energy resources but also with other "point-source" natural resources like diamonds, timber, or plantation crops.[93] However, only one type of resource – petroleum – has consistently correlated with less democracy, weaker institutions, and more conflict.[94] Third, the curse is by no means destiny. Norway and Malaysia are typically cited as oil- and gas-rich countries that managed to escape the resource curse. Finally, the resource curse thesis is not undisputed. Victor Menaldo, for instance, concedes the fact that resource reliance is *correlated* with numerous pathologies, but he rejects the idea that this relationship is *causal*.[95]

Economic ailments

Researchers have searched for the mechanisms behind the resource curse and several hypotheses have been put forward.[96] There are at least three economic problems associated with resource wealth that need to be overcome by resource-rich countries to avoid the resource curse. The first problem is the so-called *Dutch disease*, named after the experience of the Dutch who saw their manufacturing sector decline after they had discovered large gas fields in 1959. The explanation was that increased exports of natural gas led to an appreciation of the national currency and domestic inflation, and soon other sectors (agriculture and manufacturing) started to struggle in export markets. In essence, the primary resource being exported then crowds out other sectors of the economy, causing even more economic concentration, which is bad for long-term economic growth.[97]

A second challenge is *commodity price volatility*. As the rollercoaster oil prices of the past two decades have illustrated, commodity prices – and hence resource revenues – tend to be highly volatile. The sudden price gyrations and boom-and-bust cycles can erode budgetary discipline, decrease expenditures for public sector investment, and complicate efforts at state planning. In principle, this revenue volatility could be countered with countercyclical fiscal policies – that is, policies to set aside a fraction of the surplus during a boom (like, for instance, in a sovereign wealth fund), and draw down this surplus during a bust. While virtually all oil-rich governments acknowledge the importance of countercyclical fiscal policies, they rarely success-fully implement them.[98] When prices are high, leaders face expecta-tions that they should increase spending, embark on capital-intensive projects, and reward productive components of the economy with subsidies.[99] Public resources are then often spent on expensive but useless "white elephant" projects.[100]

A third challenge, which exacerbates the management of the previ-ous two, is the *enclave nature of extractive industries* such as oil. In many oil-rich countries, the oil industry is an isolated and autonomous sector with very limited interaction with other economic activities. Growth in a country's manufacturing sector, by contrast, will stimu-late growth in the rest of the economy because its employees will buy goods and services produced by other firms (the "employment effect"); its employees will learn skills that they can take to future jobs (the "learning-by-doing" effect); and the manufacturing companies themselves will buy goods from other companies, to use as inputs for their products (the "backward linkage" effect).[101] None of these productive links exist for the extractive sector. Oil and gas produc-tion and mining are very capital- and technology-intensive, meaning they create fewer jobs per dollar invested and require unique skills not well suited for other areas of economic activity. Oil and gas com-panies buy little inputs from local firms, but instead they buy highly specialized equipment that is often manufactured in high-income countries.[102]

Political ailments

The resource curse is also associated with a host of political patholo-gies, including the undermining of democracy, a lower quality of gov-ernment institutions, and the incidence of civil war.[103] Consider first the effect of resource wealth on *democracy*. Although some oil-rich developing countries such as Mexico and Nigeria have transitioned to democracy, they remain exceptions. The general rule seems to be that oil and democracy do not easily mix. One of the most common explanations has to do with the nature of the "rentier state" – a state that relies heavily on income from the export of natural resources, especially oil and gas.[104] This revenue is not generated by produc-tive operations in the national economy but by the natural endow-ments of the country. The reliance on external, non-tax revenues gives these states a large degree of autonomy vis-à-vis their citizenry, mainly through two effects.[105] The "taxation effect" implies that, as the government does not need to tax its citizens, there will be fewer societal demands for accountability. The "spending effect" refers to the ability of the government to use the abundant oil revenues to buy off consent through social services and patronage (massive subsidies to education, health, housing, food, and fuel), as well as to invest in repressive capacity.

Second, resource wealth can also have an effect on the *quality of institutions*, meaning the effectiveness of the government bureaucracy, the incidence of corruption, the rule of law, and, more broadly, the state's capacity to promote economic development.[106] One branch in the literature suggests that the effects of resource wealth on economic growth are conditional on the ex ante quality of state institutions. Countries with weak pre-existing institutions that experience a resource boom may be left worse off, whereas countries that have strong institutions can benefit from their oil or gas wealth.[107] Another branch claims that natural resource wealth can damage, or stunt, the beneficial development, of institutions – for example, by breeding corruption and rent-seeking.[108] The term "rent-seeking" can be understood as behavior to increase one's share of existing wealth without creating new wealth – from oil companies paying bribes to obtain exploration licenses,[109] governments using resource revenues to pay off supporters who help them stay in power (i.e., political patronage),[110] to a high number of persons seeking high-wage government jobs instead of working in the private sector (i.e., the guardian state).[111]

Third, several studies have found a link between the presence of mineral resources and civil war. Since the early 1990s, oil-producing countries have been about 50 percent more likely than other countries to have civil wars.[112] More specifically, the relationship between natural resource wealth and the onset of violent conflict resembles an inverted U-curve: as the value of resource wealth increases, the risk of conflict first rises, then falls.[113] When resource wealth reaches very high levels it becomes a stabilizing force, enabling the central government to either buy off or repress potential rebels, as in the case of the relatively stable oil kingdoms in the Middle East.[114] The location of oil reserves matters: oil is more likely to lead to conflict if it is found onshore, in relatively poor regions, populated by marginalized ethnic groups – many of these elements were in place in conflict-ridden places such as the Niger Delta in Nigeria, Colombia's Arauca region, and Yemen's Marib governorate. Oil wealth can spark or strengthen separatist conflict, as in the case of South Sudan (which became formally independent in 2011), Indonesia's Aceh region, or Iraqi Kurdistan.[115] Oil extraction has also financed the operations of several rebellions and terrorist groups, either through looting oil directly or engaging in kidnapping and extortion of oil companies and their workers.[116]

Notwithstanding the fact that the causal mechanisms of the resource curse continue to be hotly debated, several policy solutions have been suggested and implemented.[117] Some exporters have saved for the rainy days through stabilization funds.[118] Norway's Oil Fund is frequently touted as a model, yet it is also criticized because its investments are subject to political influence.[119] Other proposed "cures" have to do with the optimal design of petroleum contracts, currency interventions, and ways to use oil earnings for investment rather than consumption.[120] An important part of the answer to the resource curse has been to try and bring more transparency to the sector. The Extractive Industries Transparency Initiative (EITI) is a key example of this. It is a voluntary standard that is adopted by more than 50 countries. Under the EITI, companies are required to publish what they pay to governments and governments are required to report how much they earn. This transparency is seen as the best antidote to embezzlement and corruption.[121]

The future of petrostates

Oil-exporting states find themselves increasingly stuck between a rock and a hard place as two profound shifts in technology and markets are dramatically changing the oil market. In the short term, traditional producers will feel persistent pressure from the shale revolution, the unexpected increase and resilience of unconventional oil and gas production, which has contributed to a global glut and kept prices in check. In the medium term, the industry must confront a structural slowdown and eventual peak in demand owing to innovation and evolving consumer preferences, related in part to concerns over climate change. Both trends make it highly unlikely that all of the world's oil will ever be consumed and that the price of oil will rise over time.[122]

This could be a game-changer for petrostates which, for decades, have lived under the basic assumption of Hotelling's rule of optimal extraction of exhaustible resources: the owner of oil can leave the resource in the ground as a physical asset or sell it and invest the proceeds in the financial markets.[123] With oil consumption growth slowing down, and potentially reversing over time, producers might find out some day that oil under the ground is less valuable than oil produced and sold in the short term. This might encourage them to pump as much as they can, as fast as they can, to serve a shrinking market.[124] Such a strategy, however, is at best a short-term strategy to finance exploding deficits which does nothing to improve long-term growth and development prospects.[125] Only a few producers can ramp up production in the short term and most will probably shy away from increased investment in exploration, which is typically made in anticipation of rising, not falling, demand.[126]

Of course, in theory, the loss of such oil rents could be good news for petrostates, as it could make the resource curse go into reverse.[127] The problem, however, is that the transition is likely to be painful. Very few of the major oil exporters have followed the rule of Hartwick, which states that subsoil assets (e.g., oil reserves) should be transformed into reproducible surface assets (e.g., human or physical capital) to preserve long-term per capita consumption.[128] According to a World Bank study, in the period from 1980 to 2005, of the countries that got more than 10 percent of their GDP from oil and gas rents, only Malaysia has followed the Hartwick rule and diversified its economy.[129] Oil-producing countries such as Venezuela, Nigeria, Brazil, and Russia have already experienced severe economic distress and political turmoil, related in part to the decline in oil revenues after 2014.[130]

The vulnerability of major oil producers to lower oil revenues is a function of their oil-rent dependence, demography, economic base, history, and political leadership – some of these indicators are summarized in table 4.1. Saudi Arabia faces a difficult transition in the long run. It has some of the highest dependency rates on oil rents, in the order of 37 percent of its GDP in the period 2012–15.[131] Keeping its young and growing population content will be a major challenge for the Al-Saud dynasty if oil revenues remain low for a long period. Iran, by contrast, has a much broader economic base, a longer tradition of trading, and lower fertility rates. Like Iraq, its oil production is well below its potential owing to years of sanctions, which, ironically, might turn out to be an advantage in the long run. Russia is also a major loser from the shift away from oil. Even though it is less dependent on oil revenues than Saudi Arabia and some of the smaller Gulf states, its endemic corruption, autocratic tendencies, and lack of a globally competitive industrial base will leave the Russian economy in a precarious state when oil revenues dry up.[132]

Most major oil exporters recognize the need to "sow the oil" and diversify their economies. However, it is difficult to find successful models of diversification.[133] Malaysia and Indonesia have

Table 4.1 Key economic indicators of top 20 oil exporters

Country	Oil exports (mb/d)	Oil rent (% of GDP)	Oil income per capita	OPEC member?
Saudi Arabia	7.38	37.67	7,800	Yes
Russia	4.78	8.46	2,080	No
Iraq	2.83	39.87	1,780	Yes
UAE	2.50	20.36	14,100	Yes
Canada	2.23	1.16	2,530	No
Nigeria	2.11	9.43	370	Yes
Kuwait	2.04	51.95	19,500	Yes
Venezuela	1.81	14.47	2,130	Yes
Angola	1.66	28.39	2,400	Yes
Iran	1.49	18.44	1,600	Yes
Kazakhstan	1.38	10.83	2,370	No
Mexico	1.27	3.62	610	No
Norway	1.62	5.24	13,810	No
Oman	0.82	34.38	7,950	No
Algeria	0.70	15.24	1,780	Yes
Azerbaijan	0.68	20.62	2,950	No
Colombia	0.67	4.58	430	No
Brazil	0.60	1.56	240	No
UK	0.60	0.50	150	No
Qatar	0.56	10.55	24,940	Yes

Notes: Oil exports are the averages for the period 2012–16. Oil rent figures are the averages for 2012–15, except for Venezuela (2012–13 average) and Iran (2012–14 average). Oil income per capita shows the estimated value of oil and gas produced per capita in 2009.

Sources: OPEC (2017) *Annual Statistical Bulletin*; BP (2017) *Statistical Review of World Energy*; Ross, M. L. (2012) *The Oil Curse*. Princeton, NJ: Princeton University Press.

successfully diversified as manufacturers, while Dubai has attracted foreign investment in infrastructure, services, and business thanks to the creation of a massive special economic zone. It is doubtful that these experiences can simply be copy-pasted by other large oil exporters. On a per capita basis, Malaysia and Indonesia never produced as much oil and gas as other major producers.[134] The Dubai model of development has also been described as "*sui generis*" because the country so heavily depends on expatriate labor and skills, with nationals constituting only 10 percent of the population.[135]

Conclusion

This chapter has demonstrated the extent to which the energy sector – and especially petroleum – has a significant impact on the global economy and on the distribution of wealth among and within countries. Approximately $2 billion a day of petroleum is traded worldwide, making oil the largest single item in the balance of payments and exchanges between nations.[136] The tentacles of the global energy system entwine hundreds of millions of workers, some of the world's biggest companies, and contribute significantly to global GDP.

It would be myopic, however, to look at global energy markets and investments through an economic lens without appreciating the critical role played by security, environmental, and social considerations. Energy markets have never operated fully without government interference, simply because of the strategic importance of energy supply and finance to the state. Energy investments and trade have often been entangled with broader geopolitical and foreign policy interests. And the growing environmental and social backlash against fossil fuels are weighing on the prospects of private companies, state-owned corporations, and petrostates alike. It is to these environmental and social implications of our energy system that we turn next.

5

Energy and the environment: Wrecking the planet?

The world's accelerating use of energy since the dawn of the Industrial Revolution has brought huge benefits to the majority of us, in terms of higher standards of living, more food production, and a higher life expectancy. At the same time, the way in which energy is produced and consumed has caused environmental damage, the most urgent and dire of which is undoubtedly climate change. Human combustion of fossil fuels is the chief source for the emissions of CO_2 and other heat-trapping gases in the atmosphere, which is causing our climate to change at a potentially devastating pace. Clearly, climate change is much more than an issue of pollution that only environmentalists care about. It is also increasingly recognized as a threat to economic development and to global peace and security.

This chapter proceeds in three parts. First, it surveys some of the most acute environmental impacts of our current global energy system. Second, it addresses the question of whether the world is running out of fossil fuels. A final section surveys the history and key features of global climate policy, from the United Nations Framework Convention on Climate Change (UNFCCC), over the Kyoto Protocol and the Paris Agreement, to the rising "groundswell" of climate action from city mayors, companies, and other sub-state and non-state actors.

The environmental impacts of energy

This section discusses key "outputs" of our global energy system such as waste, pollution, and environmental degradation. The global energy system's wastefulness and inefficiency leads to environmental damage across at least four dimensions: climate change, air pollution, water quality and availability, and land use change (which is basically a code name for deforestation). We will discuss each in turn, but it is important to note that these are not the only environmental impacts of energy.

Climate change

The most urgent and dire output from the global energy system is the emission of carbon dioxide and other greenhouse gases into the atmosphere. It is impossible to tackle climate change without changing our energy system. Two-thirds of all greenhouse gas emissions and up to 80 percent of all CO_2 emissions stem from the energy sector. The energy sector refers chiefly to the combustion of fossil fuels, along with some fugitive emissions of carbon dioxide (CO_2) and methane (CH_4) which can occur during extraction.[1] Climate change, in essence, *is* an energy problem.

Figure 5.1 shows the geographic origins of greenhouse gas emissions. Longstanding industrialized countries no longer make up the bulk of emissions. As panel a shows, China is by far the largest emitter of CO_2, having surpassed the US in 2006. It now emits much more than the US and the EU, which are ranked second and third. India, meanwhile, has surpassed Russia and Japan to become the world's fourth largest emitter of CO_2. Panel b, which depicts emissions *per capita*, shows a different picture. Here, the US still ranks first, ahead of Russia and Japan. It is quite remarkable, however, that China (with its population of over 1 billion people) has already surpassed the EU (about 500 million people) in terms of per capita emissions. It is important to note that each country's emissions profile is different, with some (such as the United States and China) coming primarily from energy and transport, and others (such as Brazil and Indonesia) coming from deforestation and changes in land use.

Since the Industrial Revolution, humans have been adding more carbon dioxide to the atmosphere than plants and oceans can take up. As a result, CO_2 concentrations, usually measured in parts per million, have been steadily rising. There have been continuous measurements at the Mauna Loa Observatory in Hawaii since 1958, when Charles David Keeling developed a way to measure CO_2 in the atmosphere. The record of measurements is now known as the "Keeling Curve" (figure 5.2). The atmospheric concentration of CO_2 reached 400 ppm in 2016, which is considered its highest level in 3 million years.[2] The problem is that, if the concentration of CO_2

Panel a. Absolute CO$_2$ emissions

(b)   *Panel b. CO$_2$ emissions per capita*

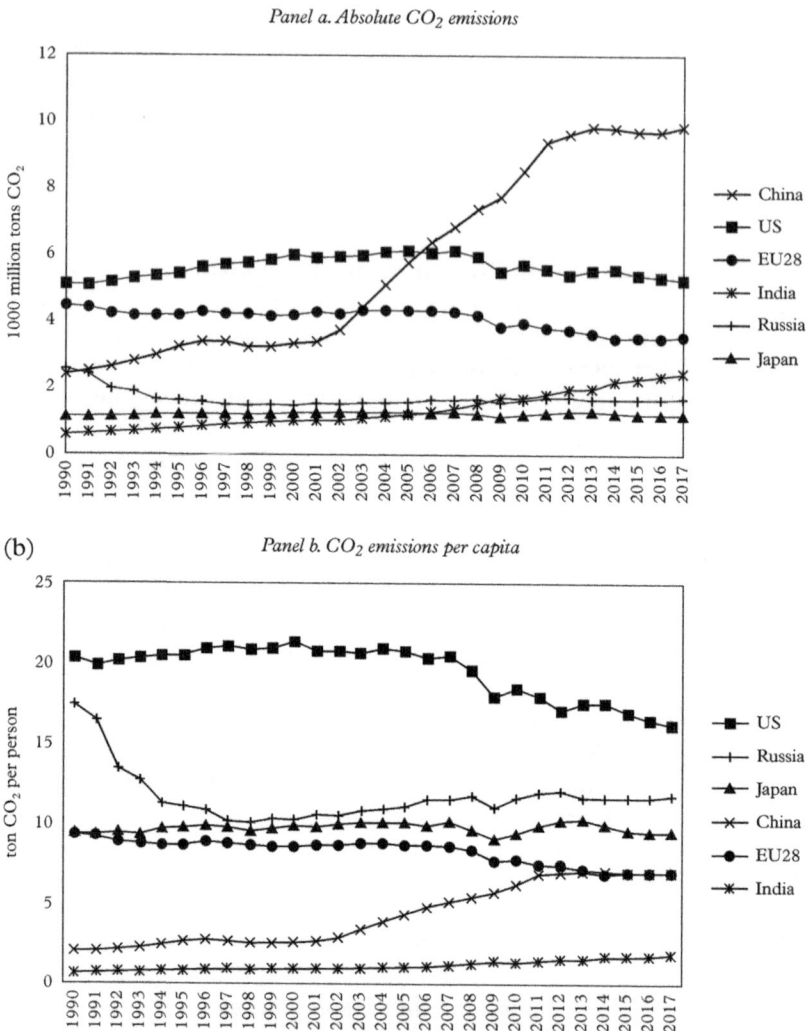

Figure 5.1   CO$_2$ emissions for selected countries since 1990

Sources: Boden, T. A., Marland, G., & Andres, R. J. (2017) Global, Regional, and National Fossil-Fuel CO$_2$ Emissions. Carbon Dioxide Information Analysis Center, Oak Ridge National Laboratory, US Department of Energy, Oak Ridge, TN, USA. Available at http://cdiac.ess-dive.lbl.gov/trends/emis/meth_reg.html; UNFCCC (2018). National Inventory Submissions 2018. United Nations Framework Convention on Climate Change. Available at http://unfccc.int/process/transparency-and-reporting/reporting-and-review-under-the-convention/greenhouse-gas-inventories-annex-i-parties/national-inventory-submissions-2018; BP (2018) *Statistical Review of World Energy.* Available at: http://www.bp.com/en/global/corporate/energy-economics.html.

Atmospheric CO_2 at Mauna Loa Observatory

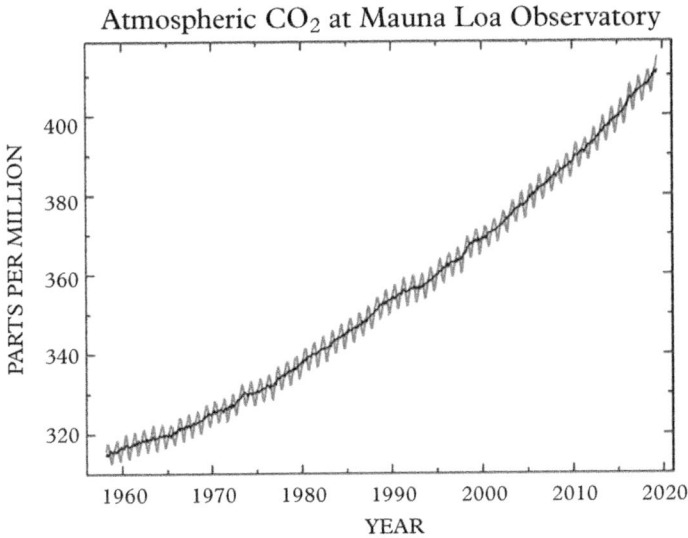

Figure 5.2   The Keeling curve

Source: NOO Earth System Research Laboratory, Carbon Dioxide Information Analysis Center.

and other greenhouse gases grows too large, too much heat will be trapped in the atmosphere. This raises the possibility of a ferocious change in climate, which will drastically affect life on the planet.

So far, since the Industrial Revolution our emissions have raised the temperature on the planet just about 1°C above pre-industrial levels.[3] In the 2015 Paris Agreement, countries agreed to limit global warming to 2°C and to pursue efforts to keep it below 1.5°C. However, with currently pledged climate policies, we are heading toward a global warming of about 3°C by 2100, with warming continuing afterwards.[4] One or half a degree may not sound like much, but as the latest report from the Intergovernmental Panel on Climate Change (IPCC) has made clear, half a degree makes a world of difference (see figure 5.3). Half a degree of warming extra could expose tens of millions more people worldwide to life-threatening heatwaves, water shortages, and coastal flooding. Choosing between 1.5°C or 2°C is a choice between a world with coral reefs and Arctic summer sea ice and a world without them.

Moreover, there are several tipping points that, once crossed, could alter the state of the system. For example, the melting of the Arctic exposes dark ocean waters that reflect sunlight much less efficiently,

Population exposed
to water scarcity

1,5°C

2°C

0   200   400   600

million people

Population exposed
to severe heatwaves

1,5°C

2°C

0   20   40

% of global pop.

Population exposed
to sea level rise

1,5°C

2°C

60   70   80   90

million people

Species loss (>50%):
plants

1,5°C

2°C

0%   10%   20%

% of plants

Species loss (>50%):
vertebrates

1,5°C

2°C

0%   5%   10%

% of vertebrates

Species loss (>50%):
insects

1,5°C

2°C

0%   10%   20%

% of insects

Further decline in
coral reefs

1,5°C

2°C

0%   50%   100%

decline in coral reefs

Ice-free Arctic
summers

1,5°C

2°C

0%   10%   20%

probab. (any yr)

Decline in marine
fisheries

1,5°C

2°C

0   2   4

million tonnes

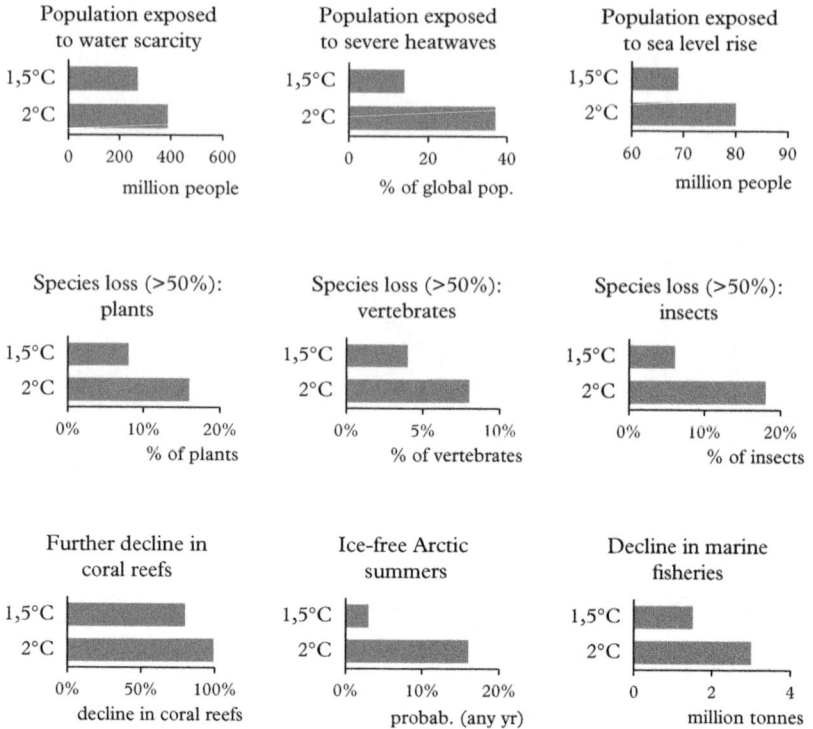

Figure 5.3   Impacts of 1.5°C and 2°C of global warming
Source: Compiled with data from IPCC (2018).

thus speeding up warming. The gradual melting of the permafrost in Siberia releases more carbon dioxide and methane into the atmosphere, which creates a feedback loop for even more warming. The disintegration of the vast ice sheets on top of Greenland or West Antarctica could commit the world to many more feet of sea level rise for centuries to come.[5] An abrupt domino effect may tip the planet into a permanent "hothouse" state if global temperatures rise by more than 2°C.[6]

Air pollution

Pollution is making the air dangerous to breathe in many cities around the world, from New Delhi to Beijing and Paris. The World Health Organization (WHO) estimates that nine out of every ten people in the world breathe polluted air that is hazardous to health and wellbeing, and that air pollution kills 7 million people every year, making it the fourth largest cause of death.[7] Around 18,000 people die *each day* as resulted of breathing polluted air, which is much greater than the number of deaths from HIV/AIDS, tuberculosis, and road injuries combined.

The energy sector is by far the largest source of air pollution: 85 percent of particulate matter (PM) and almost all of the sulfur oxides (SO_x) and nitrogen oxides (NO_x) stem from our energy production and use. These three pollutants are responsible for the most widespread impacts of air pollution, either directly or once transformed into other pollutants via chemical reactions in the atmosphere.[8]

Coal is responsible for around 60 percent of global combustion-related sulfur dioxide (SO_2) emissions. Sulfur dioxide is a pollutant responsible for lake- and forest-damaging acid rain and is a precursor to the development of small particles that damage human health. Fuels used for transport, first and foremost diesel, generate more than half the nitrogen oxides (NO_x) emitted globally, which can trigger respiratory problems and the formation of other hazardous particles and pollutants, including ozone. Cities can easily become pollution hotspots, as they concentrate people and economic activities. The impact of vehicle emissions in urban areas is heightened by the fact that they are discharged not from the top of tall chimneys but directly into the street-level air that pedestrians breathe.[9]

Particulate matter is not a specific pollutant itself, but instead is a mixture of fine particles of harmful pollutants such as soot, acid droplets, and metals. The most widely studied are those particles with an aerodynamic diameter of 10 microns or less (PM_{10}) and those with a diameter of 2.5 microns or less ($PM_{2.5}$). Inhalation of PM is strongly associated with heart disease and chronic lung disease. Since microscopic solids or liquid droplets are so small, they can get deep into the body and cause serious health problems. Numerous scientific studies have linked PM exposure to strokes, heart disease, lung cancer, and both chronic and acute respiratory diseases, including asthma.[10]

A completely separate but equally serious type of air pollution is indoors, rather than outdoors, where billions of homes rely on the burning of firewood, dung, charcoal, kerosene, and coal for warmth and heat. As the WHO explains:

> The inefficient burning of solid fuels on an open fire or traditional stove indoors creates a dangerous cocktail of not only hundreds of pollutants, primarily carbon monoxide and small particles, but also nitrogen

oxides, benzene, butadiene, formaldehyde, polyaromatic hydrocarbons and many other health-damaging chemicals.[11]

There is both a dangerous spatial and temporal dimension to such pollution. Spatially, it is concentrated in small rooms and kitchens rather than outdoors, meaning that many homes have exposure levels to harmful pollutants 60 times the rate acceptable outdoors in city centers in North America and Europe.[12] Temporally, this pollution from stoves is released at precisely the same times when people are present cooking, eating, or sleeping, with women typically spending three to seven hours a day in the kitchen.[13]

Water use and contamination

The degradation of water quality and availability is a third serious effect of our energy systems. For instance, thermoelectric power plants running on coal, natural gas, oil, and uranium require immense amounts of water to cool the combustion process, and more than 50 countries rely primarily on hydroelectric dams to generate power in their electricity sectors. These collective power plants withdraw trillions of gallons of water from rivers and streams, consume billions of gallons of water from local aquifers and lakes, and contaminate water supplies at various parts of their fuel cycle.

A conventional 500 MW coal plant, for instance, consumes about 26,498 liters of water per minute, or the equivalent of 17 Olympic-sized swimming pools every day.[14] Oil and gas production facilities, refineries, ethanol distilleries, and manufacturing firms also rely on prodigious amounts of water to transform raw commodities into usable energy fuels, services, and products. Globally, energy is the second largest user of fresh water, accounting for nearly 15 percent of global water use.[15]

Land use and deforestation

The production of energy can negatively affect land in many ways, from converting forests into plantations for energy crops, to access roads to dams and oil and gas facilities opening up areas to deforestation. One estimate suggests that 15 percent of land use changes are caused by the clearing of forests for fuel-wood and for energy crop plantations.[16]

Much of the world's farming, livestock production, and changes in land use have taken place in former forests and tropical forests (sometimes called "land use and land-use change and forestry," or LULUCF). This transformation of forestland into other uses is problematic because it directly releases carbon into the atmosphere. Forests cover about 30 percent of global land area, storing 683 billion tons of carbon, more than the total amount of carbon contained in the atmosphere. Through the carbon cycle, forests remove an additional three billion tons of carbon dioxide each year through growth, and they also absorb about 30 percent of carbon dioxide emissions from fossil fuel combustion.[17] Put another way, forests store about 45 percent of terrestrial carbon and can sequester additional carbon dioxide emissions directly out of the air. As forests grow, they sink carbon dioxide from the atmosphere in their roots, branches, and leaves (their "biomass"), in essence making CO_2 a part of the natural landscape.[18] Yet when forests are cleared, harvested, or catch fire, their stored carbon is emitted back into the atmosphere. About 36 percent of the carbon added to the atmosphere from 1850 to 2000 came from the elimination and conversion of forests.[19]

Yet the rate of deforestation worldwide averaged 13 million hectares a year between 1990 and 2005.[20] Indonesia and Brazil account for about half the emissions from deforestation, which also explains why they are (respectively) the third and fourth largest emitters of GHGs overall, behind China and the US.[21]

Other pollution

Tragically, these four major outputs – carbon, air pollution, water pollution, and the degradation of land – are not the global energy system's only environmental impacts, just the most significant. There are various other pollutants related to energy production and use, including biologically and climatologically active elements and compounds such as lead, cadmium, methane, and mercury, each deserving of more detail than we are able to provide here.[22]

Are we running out of fossil fuels?

The world economy relies on energy sources that do not only produce harmful environmental and health effects. The majority of energy sources that we rely on, fossil fuels, are also finite. This has led to periodic fears of energy scarcity.[23] These fears have not just centered on oil, but on other energy sources too. In his 1865 book

The coal question, English economist William Stanley Jevons warned of the exhaustion of economically viable sources of coal.[24] Around a century later, an American geologist, M. King Hubbert, made similar observations regarding oil. His work led the basis of the so-called "peak oil" theory.

Here, three views will be presented: the scarcity view (peak oil theory), the abundance view (emphasizing the revolution in unconventionals), and finally the excess view (taking into account the carbon budget and the risk of "stranded assets").

The scarcity view: "peak oil"

There has been a longstanding view, dating back to at least the 1970s, that the supply of oil and other resources will or could soon "peak." The notion of a supply-side "peak" refers to a situation where the rate of extraction of a resource reaches a plateau after which the rate of production enters a terminal, irreversible decline.[25] Proponents of this view argue that global statistics of energy reserves and production speak for themselves, and point to a declining availability of high-quality oil, coal, natural gas, and uranium reserves.[26]

This trend of peaking, though contested, is perhaps best exemplified by crude oil. Figure 5.4, for instance, was published in 2008 and illustrates the dominant thinking back then: the United States entered a peak for its domestic oil production in 1970, after which production entered a declining "off-ramp."[27] Moreover, panel b in the figure also indicates how, since the early 1980s, global discoveries of new oil fields have not kept pace with global production and consumption. The implication is that the best and largest reserves of oil have already been discovered and partially exhausted, and that oil is getting, as a whole, more expensive to discover and produce. The IEA added to the fears about a looming supply crunch in its 2008 *World Energy Outlook* where it highlighted the need to find the equivalent of six Saudi Arabias before 2030 to replace the existing oil fields, which experience natural declines, and to meet the growth in demand.[28]

The abundance view: "unconventionals revolution"

In recent years, fears of peak oil have largely been stymied thanks to the shale revolution. After remaining flat for around a decade, US natural gas production suddenly took off in 2007. By 2009, the United States had overtaken Russia as the world's largest gas

(a)

US Oil Production 1900 to 2050

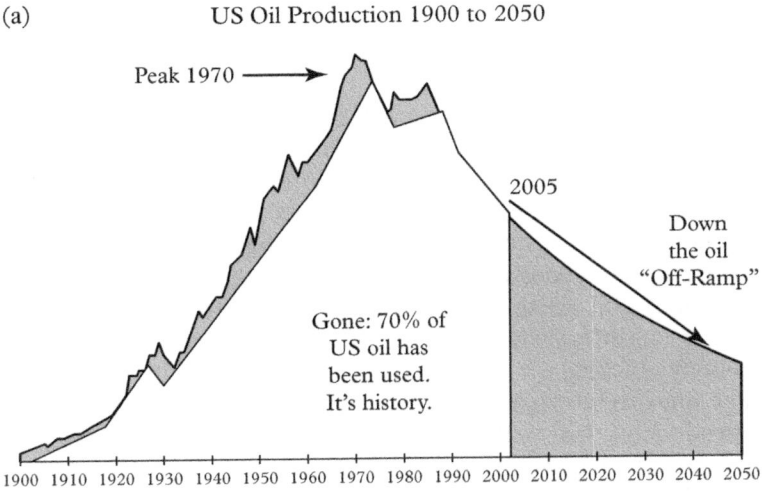

Peak 1970 ⟶

2005

Down
the oil
"Off-Ramp"

Gone: 70% of
US oil has
been used.
It's history.

1900 1910 1920 1930 1940 1950 1960 1970 1980 1990 2000 2010 2020 2030 2040 2050

(b)

Worldwide Oil Production

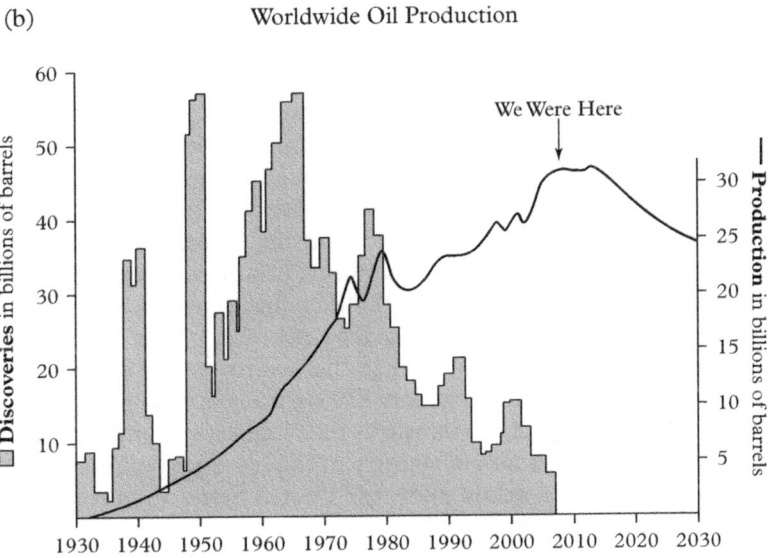

We Were Here

Figure 5.4 Trends in US and global oil production: (a) US oil
production, 1900 to 2050; (b) Global oil discoveries and production, 1930
to 2030

Source: ASPO Newsletter No. 77, May 2007, https://aspoireland.files.wordpress.
com/2009/12/newsletter77_200705.pdf.

producer. In the same year, fracking also pushed US oil production up again after a decade-long decline. In July 2018, the US produced almost 11 mb/d, an all-time high.[29]

It was a combination of three techniques, each of which had been developed separately, that allowed companies to free the gas and oil molecules trapped in layers of shale rock. They are: horizontal drilling, seismic imaging, and the ability to use hydraulic pressure to split, or "fracture," the rock to release trapped hydrocarbons. The unique mineral ownership system of the United States, whereby the subsurface rights are in private hands for private gain, served as a key enabling condition, which allowed the fracking revolution to develop extremely quickly.

The unexpected shale boom was part of a broader surge in oil and gas production. Just as the oil shocks of the 1970s had ushered in the development of new oil provinces, notably the North Sea and the Gulf of Mexico, so the oil price rises after 2000 led to an increase in exploration spending. This flurry of exploration activity has resulted in major discoveries of new conventional oil and gas fields. The increased spending on exploration also shifted the frontiers of drilling, notably to the deeper offshore, the Arctic and the Canadian tar sands.

The shale industry has been remarkably resilient in the face of falling prices since mid-2014. Even though some shale companies have gone bankrupt, and most of them have very high levels of debt, overall the sector has succeeded in cutting down costs thanks to technological innovation and streamlined production processes. The productivity gains are related to the fact that fracking is more akin to a standardized, manufacturing-type process, unlike the one-off, large-scale engineering projects that characterize many conventional oil enterprises.[30] The IEA expects US tight oil output to grow by 8 mb/d from 2010 to 2025, making this the "highest sustained period of oil output growth by a single country in the history of oil markets."[31] Growth on that scale would allow the United States to become a net oil exporter from the late 2020s.

The excess view: "unburnable carbon"

Today, one could say that the sentiment has shifted again. Now, whenever discussing the atmospheric capacity to store carbon dioxide, it's important to think about a carbon budget of one trillion tons of carbon. That is the ultimate ceiling on what scientists say we can safely emit in total.[32] Right now, humanity is emitting about 50 billion tons of carbon dioxide equivalent per year. Between 1751

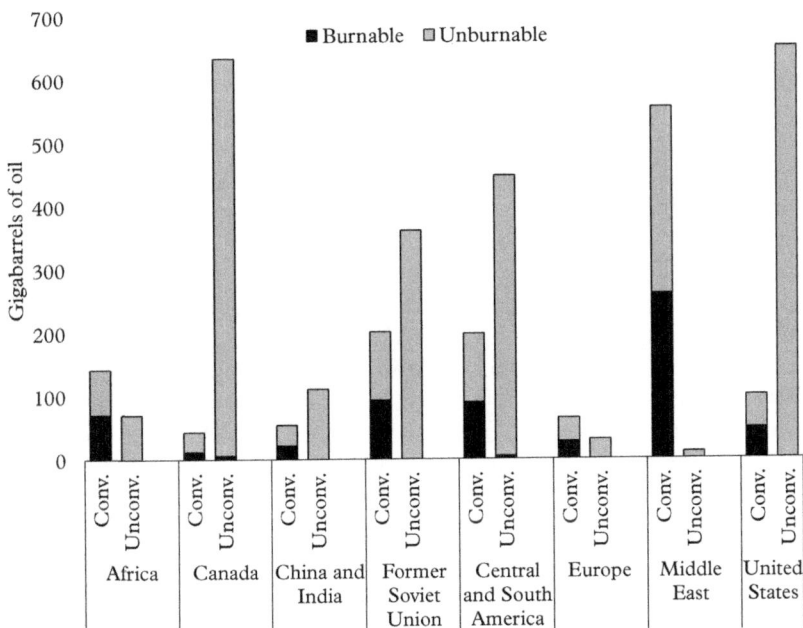

Figure 5.5   How much oil is unburnable in a 2°C scenario before 2050?

Note: "Conv." stands for conventional and "Unconv." for unconventional.

Source: Compiled with data from McGlade, C., & Ekins, P. (2015) The geographical distribution of fossil fuels unused when limiting global warming to 2°C. *Nature*, 517 (7533).

and 2012, we have emitted a total of about 435 billion tons of carbon. That means we have roughly 565 tons left that we can emit, or 40 years of emissions at today's rate.

Here's the rub: *five times* that amount of carbon – an agonizing 2,795 billion tons – is currently contained in all the proven coal, oil, and gas reserves of fossil-fuel companies and countries. Essentially, this is the fossil fuel we're planning on burning, but it will completely blow our budget.[33] Worse still, those 2,795 billion tons of carbon are worth about $27 trillion in today's (undiscounted) money, which means writing off 80 percent of that is asking fossil fuel companies to abandon $20 trillion in assets – raising concern about stranded assets and a financial carbon bubble.[34]

To have a 66 percent chance of staying within the 2°C bounds, around 30 percent of global oil reserves are deemed "unburnable" by 2050, even if one assumes widespread adoption of carbon capture and storage (CCS).[35] Figure 5.5 shows the geographical distribution of unburnable conventional and unconventional oil reserves up to 2050 in a 2°C scenario with CCS. This study suffers from restricting the period to 2050. What really matters is that post-2011 cumulative emissions stay below 270 billion tons of carbon *forever*.[36] Still, the basic insight that the bulk of oil reserves should stay in the ground remains valid. According to this study, which took into account various lifting costs for oil, Canada should not touch any of its tar sands, the US should leave its tight oil reserves in the ground, while the Arctic should be left unexploited.

Ultimately, these carbon constraints mean that "peak oil demand" must precede "peak oil supply." That would be quite a reversal of fortunes for the oil industry. Historically, oil demand has grown in step with economic output. Since 1965, oil consumption has risen from about 30 mb/d to close to 100 mb/d in 2019. Most scenarios project world oil demand to continue growing for the next two decades. The IEA's "New Policies" scenario, for instance, foresees an increase of 10 percent in world oil demand from 2016 to 2040; and under a "business-as-usual" scenario, the IEA even forecasts an increase in world oil demand of 25 percent. The energy outlooks of OPEC and international oil companies all project similar demand growth. At the same time, several trends are indicating that oil use may be heading for a peak (see box 5.1).

Box 5.1   Peak oil demand

There are big market, technology, and policy shifts afoot in the oil market. They include a slowdown in oil consumption growth, improvements in alternative energy, and the carbon constraints imposed by the threat of climate change. Taken together, they may represent the biggest change in the oil market since the introduction of the mass-produced car with an internal combustion engine more than 100 years ago.

In all projections, future oil demand growth hinges on a small number of countries – about 80 percent of incremental demand comes from China and India alone – and a shrinking set of sectors – petrochemicals, aviation, shipping, and freight – where growth is expected to offset declines in all, and we mean *all*, other countries and sectors. Even if these key markets grow as assumed, technological and market shifts might start to challenge the dominant position of oil much more quickly than many assume.

The transport sector, oil's key stronghold, is widely believed to be on the cusp of a major transformation, with the advent of electric vehicles (EVs), automated driving, and ride-sharing. The full impact of such changes remains to be seen – for example, the use of self-driving cars might lead to more miles logged, which might increase demand for fuels – but the potential for disruptive change is there, especially if the three models are combined. Efficiency improvements in internal combustion engines could take a bite out of oil demand even before EVs come into the picture, and "modal shifts" (to non-car-based transport like public transport or electric bicycles) should be added into the equation as well.

Oil demand in the petrochemicals sector is vulnerable too,[37] as its products (e.g., plastics, cosmetics, fertilizers, and synthetic cloths) can also be manufactured from natural gas and biomass-based feedstock. There is also a growing backlash against (single-use) plastics following revelations that there will be more plastic than fish in the world's oceans by 2050, and that microplastics are finding their way into the food chain, posing significant threats to health.[38]

Government policies to mitigate the financial and environmental costs of oil consumption can also make a huge dent in global oil demand. A prime example is the recent curtailing of oil subsidies in countries such as China, India, and Indonesia, the main engines of oil demand growth in Asia over the coming years.[39] Policies to combat air pollution and mitigate climate change could also favor a transition away from oil, and diesel in particular. France, the Netherlands, and the United Kingdom have recently announced a complete phase-out of conventional combustion engines over the next two decades, while China and India have declared their intent to follow this example.[40] Security of supply concerns, coupled with a desire to become a forerunner in the clean energy race – the commercial race to become a leading supplier of clean energy technology – could further erode the status of oil in countries such as India and China.

Together, structural changes in technology and markets are slowing down the demand for oil, which is heading for a peak, though the date at which this will actually be reached is highly uncertain. Moreover, a demand peak could very well be followed either by a long plateau rather than a sudden plunge.

Global climate governance

Climate change can be regarded as the defining global issue of our time. Since the mid-1990s, country delegates and other stakeholders have come together in a series of meetings under the aegis of the United Nations to try and address the issue. While the international community has dealt quite successfully with other global environmental challenges, like ozone depletion, the problem of climate change has proven to be much more difficult to solve. Over time, as multilateral negotiations have proven incapable of bending the global emissions curve, other actors have increasingly stepped in. This section surveys the history and key features of global climate policy, from the UNFCCC convention, over the Kyoto Protocol and the Paris Agreement, to the rising "groundswell" of climate action from city mayors, companies, and other sub-state and non-state actors.

UNFCCC

The first real climate-focused treaty was the UNFCCC, the United Nations Framework Convention on Climate Change, adopted at the Rio Earth Summit in 1992. Its ultimate objective was to achieve the "stabilization of greenhouse gas concentrations in the atmosphere at a level that would prevent dangerous anthropogenic interference with the climate system." It is important to note that this is a political consensus – this treaty text was adopted among the close to 200 nations of the UN, and everybody had to agree to it, including Russia, OPEC, and all the developing countries. Nobody really knows what "dangerous" level of interference actually means – we will turn to that later.

Apart from the general objective, the UNFCCC also established two key principles of the global climate regime. One was the "precautionary principle," according to which the lack of full scientific certainty should not be used as an excuse to postpone action when there is a threat of serious and irreversible damage. The other was the principle of "common but differentiated responsibilities," the idea that rich countries should bear the biggest burden in terms of mitigation, both because they have historically been able to industrialize in a very polluting and carbon-intensive manner, but also because they are richer and hence have more resources to deal with it.[41] One way to operationalize such an approach was "contraction and convergence," where high-emitting, rich countries cut their emissions, but low-emitter, poor countries gradually increased their emissions to a threshold sufficient to reach healthy levels of economic and social development.

Kyoto Protocol

The UNFCCC was just a framework convention. The generic goal and principles still had to be translated into concrete action points, which led to the negotiation of the Kyoto Protocol, signed in 1997. This protocol contained concrete targets and timetables: the developed countries (those listed in Annex I of the Protocol) had to reduce their emissions by 5.2 percent below the 1990 baseline by 2012. It took a very long time for the Protocol to enter into force, basically because the US refused to join. Under the Clinton administration, the Republican-dominated Congress was very hostile to the treaty because it put carbon limits on the US economy but not on China; and after the election of George W. Bush in 2001, the US made it clear that it would never ratify it. Thanks in part to Russia's ratification in 2004, the protocol could finally enter into force in 2005, some eight years after its signing.

The Protocol introduced a couple of so-called flexible mechanisms which basically allowed developed countries to reach their emission reduction targets in ways other than mitigating greenhouse gas emissions on their own territory. Through the Clean Development Mechanism (CDM) they could invest in clean projects in *developing* countries – say, a windmill farm in Ethiopia – and earn carbon credits for it. Joint Implementation (JI) is a similar mechanism but it applies to projects in other *developed* countries (mainly countries in eastern and central Europe) – for example, sponsoring a hydro dam in Romania. Kyoto also introduced the possibility of emissions trading. All of these flexible mechanisms were introduced on US insistence but, ironically, it would be the EU who become the major champion of emissions trading.[42] The idea behind these mechanisms was that emissions should be reduced where it is most economical to do so.

The Kyoto Protocol is generally regarded as a failure. First, the Kyoto targets were too modest. The –5.2 percent target for the developed countries was a far cry from the –20 percent target suggested in the Toronto Declaration (1988) for 2005. In the years after the Kyoto Protocol was signed, the state of climate research advanced and showed that the problem was actually much worse and urgent

1997

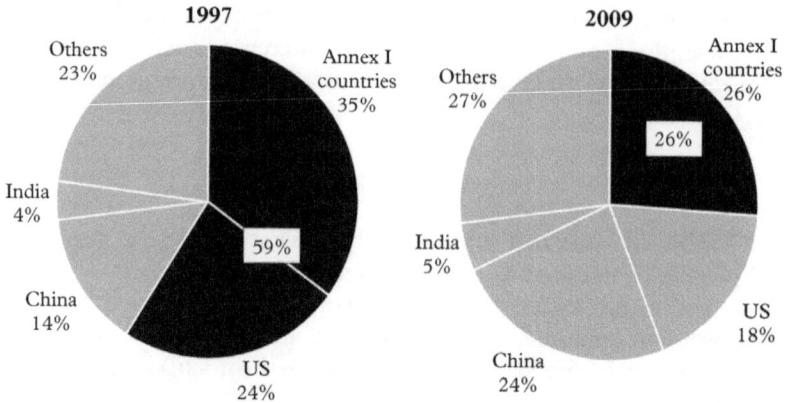

Others 23%

Annex I countries 35%

India 4%

59%

China 14%

US 24%

2009

Others 27%

Annex I countries 26%

26%

India 5%

US 18%

China 24%

Figure 5.6   Share of global CO_2 emissions under binding Kyoto targets

Sources: Boden, T. A., Marland, G., & Andres, R. J. (2017) Global, Regional, and National Fossil-Fuel CO_2 Emissions. Carbon Dioxide Information Analysis Center, Oak Ridge National Laboratory, US Department of Energy, Oak Ridge, TN, USA. Available at http://cdiac.ess-dive.lbl.gov/trends/emis/meth_reg.html; UNFCCC (2018). National Inventory Submissions 2018. United Nations Framework Convention on Climate Change. Available at http://unfccc.int/process/transparency-and-reporting/reporting-and-review-under-the-convention/greenhouse-gas-inventories-annex-i-parties/national-inventory-submissions-2018.

than previously thought, which made the −5.2 percent target even more inadequate.[43] Moreover, the choice for 1990 as the baseline year against which countries had to reduce their emissions was a big gift to countries like Russia and Ukraine, because their economies had collapsed by about 30 percent in 1991 with the fall of the Soviet Union. In practice, this meant that Russia could actually *increase* its emissions under the Kyoto Protocol. The introduction of emissions trading meant that this excess "hot air" from Russia and Ukraine could be bought by third parties.[44]

Second, the biggest emitters did not take part in the emission reduction efforts. In 1997, the year the Kyoto Protocol was adopted, countries subject to binding emission reduction targets – and this includes the US – still represented a sizeable 59 percent of global CO_2 emissions. However, by 2009, this share had fallen to a meagre 26 percent, and it kept on shrinking afterwards because of the US withdrawal and the emission increase from China and other developing countries (see figure 5.6). This explains the seemingly paradoxical situation whereby the aggregate Kyoto targets were met, yet global emissions kept on rising anyway.[45]

Third, the flexible mechanisms did not function as intended: the EU's Emissions Trading System, the first of its kind, suffered from a lot of problems in its early days: emission allowances were handed out for free to big polluters (which boiled down to heavy subsidies to those companies), the carbon price collapsed due to the economic crisis, and there were even cases of theft and fraud.[46] More than 50 percent of CDM funds went to China, and the "additionality" of the emissions reductions of many of these projects was contested. That is, it was not clear whether these investments actually resulted in *additional* greenhouse gas emission reductions compared to a baseline scenario.[47]

Finally, there were no real penalties for countries that did not meet their national greenhouse gas reduction targets. So, when Canada left the Kyoto Protocol in 2011, before the end of the first commitment period, it faced no sanctions whatsoever.

The search for a post-Kyoto agreement

As the first commitment period of Kyoto expired in 2012, the search was on for a successor treaty. In 2009, expectations ran high in the run-up to the Copenhagen summit – known as "COP15" – in part because President Obama had just taken office in the White House. The purpose of COP15 was to craft a new Kyoto-style treaty – that means, a top-down negotiated treaty with binding national targets and timetables for emissions reduction. However, at the eleventh hour, there was still no deal and the EU was barred from the room where the so-called BASIC countries (Brazil, South Africa, India, and China) and the US hammered out a deal. To many observers, this deal corresponded to a vague and meaningless text but there was no time left for a real discussion so the plenary had no choice but to "take note" of this text. The only real innovation was that it mentioned for the first time the 2°C target and the Green Climate Fund, a financial mechanism to assist developing countries in adaptation and mitigation practices to counter climate change. The Copenhagen Accord even set a goal for the world to contribute $100 billion per year by 2020.

Two years later in Durban, lacking a successor treaty to Kyoto, the climate negotiators agreed to extend the Kyoto Protocol to 2020 in a second commitment period. However, many developed countries no longer wished to take part, each for its own particular reasons: Russia was quite content with the Arctic melt because it brought new oil and gas supplies to the fore and also raised the prospect of better agricultural yields in the country; Canada was investing heavily in tar sands extraction; and Japan was struggling with the consequences of the 2011 Fukushima nuclear disaster which led it to shut down all of its reactors and burn much more fossil fuels (in particular natural gas). Belarus, Ukraine, and New Zealand also dropped out, while the US still did not take part in the Kyoto process.

The Paris Agreement

A diplomatic victory was achieved in December 2015 when, after years of fruitless negotiations in search of a real post-Kyoto agreement, the Paris Agreement was signed. In the text, the countries pledged to "Holding the increase in the global average temperature to well below 2°C above pre-industrial levels and pursuing efforts to limit the temperature increase to 1.5°C above pre-industrial levels."[48] The Agreement gathered sufficient ratifications to enter into force already in November 2016, only 11 months after its signing. This is remarkable compared to the 8 years that it took for the Kyoto Protocol to enter into force.

The Paris Agreement was designed quite differently from the Kyoto Protocol. In the run-up to Paris talks in December 2015, *all* countries – not just the developed ones – had to prepare national plans on how to reduce their emissions. Even though the principle of common but differentiated responsibilities from the UNFCCC still stands, the firewall between OECD and non-OECD countries – which had been one of the key causes of gridlock in the negotiations over a post-Kyoto deal – seems to have been breached. The "nationally determined contributions" (NDCs) are unilateral, voluntary pledges. They did not arise from high-level multilateral negotiations but rather from domestic deliberations, making climate mitigation policies a much more domestically-driven process.[49]

Table 5.1 lists some of the NDCs by major players as they were submitted to the Paris climate talks in December 2015. Note that the base year for the US (2005) is different than for the EU (1990). Emissions in 2005 were higher than those in 1990, so the US goal could be regarded as less ambitious. China's and India's targets do not relate to absolute emissions reduction, but to relative emissions reduction (relative to GDP growth). This means that China and India, under their NDCs, can still emit more greenhouse gases in 2030 than today, even as the carbon intensity of their growth has fallen.

The vital question is whether the pledges, which are voluntary, will actually be carried out. The whole process received a blow in

Table 5.1 Selected national targets for the Paris Agreement

EU (28)	US	China	India
Absolute emissions **40%** below 1990 levels by 2030.	Absolute emissions **26–28%** below 2005 levels by 2025.	Emission intensity of GDP **60–65%** below 2005 levels by 2030. CO_2 emissions peak by 2030. Non-fossil fuels to be 20% of energy supply by 2030.	Emission intensity of GDP **33–35%** below 2005 levels by 2030. Electricity-generating capacity to be 40% non-fossil fuel-based by 2030.

Source: UNFCCC.

June 2018, when US President Donald Trump announced the US withdrawal from the Paris Agreement. The process of withdrawal will only be completed by late 2020, coinciding with the next presidential elections. But in the meantime, Trump has dismantled an abundance of Obama's climate regulations, including the Clean Power Plan. Studies show, however, that US emissions could still fall despite Trump's policies. The key reason is that coal is being kicked out of the power sector due to market forces (cheap shale gas and cheap renewables).[50]

Even if the NDCs are carried out, they still lead us to overshoot the target of keeping global warming "well below 2°C." Therefore, the Paris Agreement foresees a process which is called "pledge and review." The idea is that countries will review and renew their pledges every five years, starting in 2020. The hope is that countries will raise their ambition to close the emissions gap – a process sometimes also referred to as "ratcheting up." This is a big experiment and the outcome of this process is still uncertain. What is certain is that any delay in action means that carbon emissions will have to fall at an even faster pace afterwards. To stay within the 2°C bound, CO_2 emissions need to be "net zero" by around 2070. If we overshoot our carbon budget, we will have to take carbon out of the air or cool the climate through geoengineering – but both options are very controversial (see box 5.2).

There is another way in which the 2015 Paris climate summit marks a historic turning point in the global climate regime – namely by explicitly seeking to embrace climate action by sub- and non-state actors. In addition to action at the state level, a "groundswell"

Box 5.2   Negative emissions and geoengineering: Viable options?

If we overshoot our budget, there are two uncertain but potential escape routes. One is to suck carbon out of the air – this is called "negative emissions technologies" (NETs) in IPCC parlance. The other is to engineer the climate and oceans ourselves – this is called "geoengineering."

NETs is a term for a wide range of technologies. The best-known technique for sucking carbon out of the air is actually very simple: planting trees. This is called afforestation. During their lifetime, trees consume CO_2 and store it. There are other techniques available like pouring iron into the sea but, as said, only few technologies have proven to work at large scale. However, many scenarios (e.g. of the IPCC and the IEA) rely on these NETs. Note that carbon capture and storage (CCS) is not a negative emission technology by itself. If you equip a coal-fired power plant with CCS, you can capture some but not all of the carbon. So the coal plant would still emit CO_2. CCS is only a NET if you install it on a biomass plant (because the biomass has sucked carbon out of the air prior to being combusted).

Geoengineering takes us even further into the realms of science fiction. There are a whole range of solar radiation management techniques that try to "cool" the earth artificially, e.g. by creating clouds, painting all roofs and roads in white so sunlight is reflected, and many other fancy technologies. A key risk is that this amounts to a sort of "get-out-of-jail-free card," like in the board game Monopoly. This creates a "moral hazard": research reveals that people – especially the rich and powerful (who have larger carbon footprints) – would be less inclined to cut their carbon footprints if geoengineering was an option.[51] A second risk is that you are interfering with a very complex system – the climate system – and that this may trigger unintended consequences. That is why a moratorium on geoengineering experiments was approved within the UN's biodiversity regime.

---

of climate actions has emerged in recent years as cities, regions, businesses, and civil society groups have started to step up their acts on mitigation and adaptation.[52] These trends had begun before the Paris Agreement was signed,[53] but they have since gained prominence. For instance the C40 Group now connects 94 of the world's largest cities to take bold climate action and collectively covers one-tenth of the world's population.[54] Subnational emissions trading systems have proliferated, with trading in California, Quebec, Ontario, and Chinese municipalities coming online in the last few years.[55] The RE100 initiative brings together 179 of the world's biggest companies – including global brands such as Apple, Facebook, IKEA, General Motors, Microsoft, Google, Nestlé, Unilever, and Starbucks – committing to source their power from 100 percent renewable electricity.[56]

Conclusion

From climate change to Beijing's "airpocalypse" and the "plastic soup" and twirling vortexes of trash in the Pacific Ocean, the environmental consequences of fossil fuels are becoming more and more acute. For a long time, states have skirted away from directly targeting fossil fuels. The Kyoto Protocol focused on a basket of six greenhouse gases (not just CO_2) and also included carbon sinks (afforestation and reforestation) to give countries flexibility to decide in which sector they wanted to take action, and give them the option to leave the energy sector largely intact. The text of the Paris Agreement still did not mention the words "coal," "oil," or "fossil fuels."

The problem, however, is that climate change is not like any other environmental problem. There is no way to decarbonize the economy without tackling head on the production and consumption of fossil fuels. Countries have found it relatively easy to agree to global bans or phase outs of ozone-depleting substances (under the Montreal Protocol), mercury (under the Minamata Convention), and whaling (under the International Whaling Commission), but it is an entirely different story to phase out fossil fuels which still provide more than 80 percent of worldwide primary energy. Still, slowly but surely, international norms appear to be shifting,[57] and many of them focus on the supply-side of the climate problem.[58] A coalition of more than 30 countries, the Powering Past Coal Alliance, has agreed to phase out unabated coal-fired power plants by 2030. Academics are proposing to negotiate a Fossil-Fuel Non-Proliferation Treaty to ban fossil fuel extraction, similar to international arms control treaties.[59]

A switch away from fossil fuels and toward other, low-carbon sources of energy could bring major changes to global geopolitics and the macro-economy. As we will see in the next chapter, both our current energy system and the impending transition to renewables raise important questions with regard to social justice, fairness, and equity.

6
Energy and justice: Equitable and fair?

Although perhaps not an obvious connection, the intersections between the global energy system and issues of social justice are persistent, expansive, and sobering. For example, the costs of climate change will disproportionately fall on the weakest and least developed countries as well as the poorest in developed nations, while any benefits, if there are any, will likely accrue to the rich and powerful.[1] Some social and environmental burdens stem from not having enough energy – from lack of access to modern forms of energy services, under-consumption, and poverty. Others result from having too much energy – from waste, over-consumption, and pollution.[2] Modern energy systems can serve as a sword without a hilt: they can produce many benefits, but can also cause significant health burdens that shorten lives, undermine the conditions for happiness, and obstruct a more equitable use of energy.

Energy decisions should therefore be considered as moral ones, with moral implications. Hence, "energy justice," a burgeoning field of study among energy scholars, evaluates where injustices emerge in energy decision-making processes, which segments of society are affected (and ignored), and which procedures exist for their mediation in order to reveal and reduce these injustices.[3] It is with this appreciation for justice that the chapter examines three ways in which energy systems relate to social justice concerns: first, energy and poverty; second, the linkages between energy, democracy, and human rights; and, third, the ethics of the energy transition, including the challenge of making a "just transition" away from fossil fuels.[4]

Energy deprivation

Energy poverty

Energy lies at the heart of human development. It allows for the illumination of homes, the cooking of food, the pumping of groundwater, the refrigeration of vaccines, and many other services that are essential to achieve an adequate standard of living. Such services are often taken for granted in wealthy countries but the reality is that energy poverty is a plight that affects millions of people in developing countries. In countries such as South Sudan, the Central African Republic, Chad, and Sierra Leone, less than 10 percent of the population has access to electricity.[5] Without access to reliable and affordable energy, communities live in darkness and schools and hospitals can barely function.

Globally, roughly one billion people – mostly concentrated in sub-Saharan Africa and South Asia – live their daily lives without electricity, while hundreds of millions more live with unreliable or expensive power. At the same time, three billion people – that is almost 40 percent of the world's population – still rely on traditional, solid fuels such as firewood and animal dung for cooking and heating, resulting in indoor and outdoor air pollution that has widespread health impacts (see figure 6.1). Poor energy access affects not just individual households but also businesses. According to the World Bank, economies in sub-Saharan Africa suffered 690 hours of power outages on average in 2015 – the equivalent of a whole month of blackouts.[6]

There is now a broad international consensus that energy poverty poses a key barrier to economic development and has impacts on a wide range of sustainable development indicators, including health, education, food security, gender equality, livelihoods, and poverty reduction. While energy was still conspicuously absent from the Millennium Development Goals, it has been prominently incorporated into the Sustainable Development Goals (SDGs), which were adopted in 2015 by 193 member states of the United Nations. The growing global concern over energy is embodied in SDG 7, which aims to "Ensure access to affordable, reliable, sustainable and modern energy for all" by 2030.

In the past two decades, millions of people have gained access to electricity. Developing countries in Asia, led by China and India, have made significant progress. The spread of off-grid renewable energy systems – particularly solar – also offers a glimmer of hope

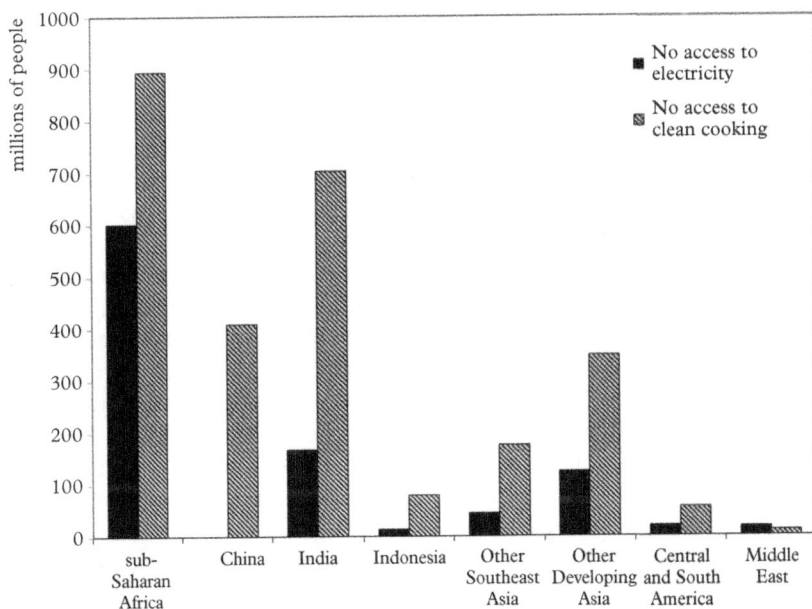

Figure 6.1   The global scale of energy poverty, 2017

Source: IEA (2018) *World Energy Outlook.* Paris: OECD/IEA. https://www.iea.org/
SDG/cooking/ and https://www.iea.org/SDG/electricity/.

(see box 6.1). Yet these efforts will need to accelerate if the world is going to meet SDG 7. The number of people gaining access to electricity in sub-Sahara Africa is outpaced by population growth. On current trends, about 674 million people, mainly in Africa, may still be without power in 2030. The goal of universal access to clean cooking remains largely elusive.[7]

Energy poverty is typically considered as a development issue but it is also a security issue. It has been described as a "threat multiplier," which means that it can exacerbate conditions thought to enable unrest and instability, such as poverty, migration, and environmental instability.[8] Without electricity to support economic growth and modernization, the pathway to jobs and the middle class for hundreds of millions of young people in the developing world will be stymied, sowing social dissatisfaction. Energy poverty can also place women and children at risk of injury and violence during fuel gathering.[9]

Box 6.1   How off-grid renewable energy solutions empower the poor

Historically, wider access to electricity was almost always achieved by extending centralized grids powered by fossil fuels. Since around 2011, however, renewable energy has become an increasingly realistic alternative due to the confluence of two major trends: cost declines and mobile money.

The cost of various renewable energy technologies, especially photovoltaic (PV) solar panels, has come down dramatically over the past decade. As a result, standalone solar systems unconnected to the main grid have become a competitive option in rural and peri-urban areas. Solar PV systems are now significantly cheaper than diesel-fired generators and kerosene lamps. Efficient home appliances (lights, fans, televisions) have also fallen in price.

Simultaneously, cell phones and mobile banking offer new ways to buy and sell energy services. In much of sub-Sahara Africa, more households have mobile phones than have access to electricity.[10] "Pay-as-you-go" solar is particularly popular. To obtain the equipment they want (typically a solar module with a battery and small appliances such as LED bulbs and mobile phone chargers),

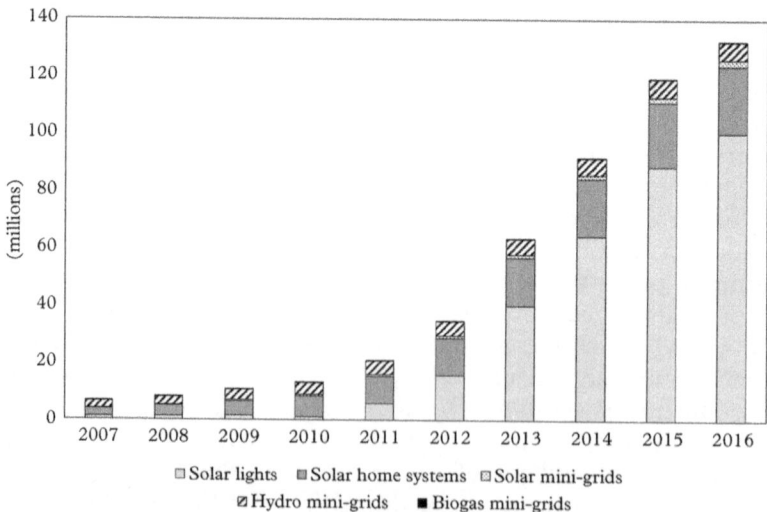

Figure 6.2   The population served by off-grid renewable energy solutions, worldwide

Source: IRENA (2018) *Off-Grid Renewable Energy Solutions.*

customers use their phones to pay a small up-front amount and then pay off the balance in weekly or monthly instalments.

A quiet energy revolution is now underway, providing light and power to households and entrepreneurs via off-grid renewable energy systems (figure 6.2). Replacing kerosene by solar lighting is only the beginning. Cheap capital and high levels of solar radiation (insolation) are combining to lift the energy available in some of the world's poorest countries. Unlike previous initiatives to electrify sub-Saharan Africa, this surge of activity is not led by states but by private companies that want to benefit from a growing market.

Fuel poverty

"Fuel poverty" is a term that refers to households that have to spend over 10 percent of their income on energy services necessary to meet basic needs.[11] While the term "energy poverty" is commonly used to refer to a *lack* of access to electricity and dependence on solid biomass fuels (e.g. wood, charcoal, and dung) for cooking and heating, the term fuel poverty is traditionally associated with energy deprivation in the developed world.[12] This framing is slightly misleading, however, as fuel poverty is found throughout the world. In recent years the issue has received increased attention in Eastern Europe, the Asia Pacific, Africa and even North America.[13] Fuel poverty today is escalating everywhere, due to the coincidence of rising fuel prices with decreasing household purchasing power; it is probably more helpful to think of it as a global problem, rather than limiting the term only to wealthy countries.

Fuel poverty typically results in inadequately heated houses (though this is not the only consequence), with a wide range of associated health impacts, including increased risk of respiratory and circulatory diseases in adults, asthma in children, thousands of excess winter deaths among the elderly, and increased risk of mental health illness and social isolation.[14] The groups most vulnerable to fuel poverty are low-income households with children, the elderly, the disabled, and people with long-term health conditions.[15]

Rising prices for fuel have a disproportionate impact on low-income families, who must pay a larger share of their income to meet their energy needs. One UK study found that households with incomes in the top 20 percent spent only 4.2 percent of their budget on fuel, while households in the bottom 20 percent spent 12.1 percent of available income.[16] The burden for low-income households was nearly three times greater than that for wealthy households. The inequity of this situation is compounded by the fact that the poor do not have the resources to make cost-saving capital investments on improving household energy efficiency. As another UK study stated, "the prospect for low-income groups is of continuing dependence on increasingly expensive electricity and gas supplies, making them an "energy underclass" at continued or increased risk of fuel poverty."[17]

Most severely, fuel poverty leads to "excess winter mortalities," quite literally killing people who go without essential heat. One 2018 report calculated 167,690 excess winter deaths – defined as the extra deaths in the four winter months in comparison with the previous and succeeding four months – from 2011 to 2017 in the UK, of which they attributed 50,310 to cold housing conditions, and of this, 16,770 to fuel poverty.[18] For comparison, that is more than the number of people who die from breast or prostate cancer each year. Interestingly, most deaths from fuel poverty occur not in the coldest areas of Europe, but in areas with milder winters such as Malta, Portugal, Cyprus, and Spain.[19] This is because colder countries tend to spend more on heat, to have better quality housing, and to have more generally accepted lifestyle adjustments to cold weather such as wearing better clothing or changing daily activities.

Energy, democracy, and human rights

A second set of risks relate to democracy and human rights. "Energy democracy" is a term that reflects "struggles around the social, economic and political relations embedded in energy."[20] One review of the growing field of study notes that energy democracy in its broadest sense identifies two meanings: it either denotes the normative goal of decarbonization and energy transformation, or it describes already existing examples of decentralized and mostly bottom-up civic energy initiatives.[21] Human rights abuses – which generally cut across the basic categories of civil, cultural, economic, political, and social rights and extend to cover issues of property, economic development, human health, safety, and the natural environment[22] – are implicated by energy systems.

Corruption, lobbying, and disinformation

The energy democracy lens reveals a series of troubling ways in which energy systems can erode democratic norms. A first threat is corruption. As we saw in chapter 4, fossil fuel-based energy generates huge profits and rents. This money can often be used by companies to bribe politicians, bypass environmental regulations, circumvent democratic procedures, and evade tax payments. Various types of corruption – the looting of oil revenues by well-placed military and political representatives, kickbacks, illegal commissions, and illicit oil-for-arms deals – are routinely associated with revenues from oil exploration and production.[23] In Iran, for example, $11 billion in oil revenues "vanished" over the course of nine months in 2010.[24] In Nigeria, an independent audit recently revealed $540 million "missing" from $1.6 billion in advance payments to oil companies to develop new fields, in addition to 3.1 million barrels of oil that remain unaccounted for, along with the disappearance of $3.8 billion in dividends from natural gas.

The corruption associated with oil funds can assume blatant forms in Africa and other parts of the developing world. But the influence of profits from the energy sector is equally pervasive in developed countries, even if this influence is exerted in less direct ways. It is no secret that the politics of climate change in the United States is heavily influenced by the fossil fuel lobby, which is determined to obstruct any attempt to legislate a limit on CO_2 emissions, like the 2009 Waxman-Markey Bill.[25] The world's five largest publicly traded oil and gas companies (ExxonMobil, Royal Dutch Shell, Chevron, BP, and Total) spent nearly $200 million a year in the period 2016–18 on lobbying designed to control, delay, or block binding climate policy, according to a 2019 report by the environmental NGO InfluenceMap.[26]

Next to lobbying, these companies have also spread false information about the urgency of climate change. Misinformation – perpetuated by fossil fuel interests – that disparages and distorts the science underpinning climate change has prompted former NASA climatologist James Hansen to argue that CEOs of energy firms that are found guilty of disseminating such information should be tried for "crimes against humanity."[27] The resources behind these attempts to mislead the public are not trivial – Robert Brulle laboriously examined the tax records of "counter climate change movements" and noted more than 91 institutions funded by 140 different foundations from 2003 to 2010.[28] Together, these 91 institutions reported an annual income of more than $900 million. Such efforts have been systematically shown to artificially polarize climate change discussions and shape policy in the United States.[29]

Even renewable sources of energy can become intertwined with corruption and antidemocratic tendencies. Lasse Eisgruber analyzed the extent to which hydropower in Laos, wind energy in Mongolia, and solar energy in the Middle East and North Africa could contribute to poor governance.[30] He found that exports of electricity from these renewable sources did lead to a crowding-out of exports, higher corruption, and lower government accountability. Walker and Baxter also note that the opposition to the development of local wind farms in Candida was highly associated with a lack of procedural justice including few opportunities to take part in siting discussions.[31]

Secrecy and authoritarianism

A second threat to democracy is how energy systems embed or rely on secrecy and authoritarianism, or marginalize community input. For example, given the potential for catastrophic accidents inherent in the use of nuclear energy, and the ongoing problem of how to dispose of radioactive waste safely, decisions regarding nuclear power ought to involve some degree of public participation.[32] Nonetheless, in the United States, India, France, and the former Soviet Union, nuclear power licensing and siting have all violated the principles of due process and informed consent in various ways. Legally unable to prohibit the public from participating in administrative procedures, the nuclear industry in the United States has made use of a number of different tactics to minimize public participation in decisions regarding licensing and permitting. Some of the more common practices have included:

- setting the timing of hearings so that interveners and the public must gather information and present safety concerns before receiving the documents necessary to adequately evaluate the risks of a project;
- interpreting rules narrowly to exclude unfavorable evidence and safety concerns; and
- attenuating inquiries into safety issues by prematurely ending proceedings.

These practices were designed to prevent the public from having a significant influence over the siting of nuclear facilities and infrastructure. Public involvement in nuclear hearings was discouraged, and when it did occur the process often alienated the public and served to legitimate nuclear power as a foregone conclusion.[33] MIT nuclear engineering professor David J. Rose once quipped that discussing concerns about nuclear power with members of the nuclear community had "the intelligence, grace, and charity of a duel in the dark with chainsaws."[34] Nuclear power projects in India have been subjected to expedited environmental impact assessment (EIA) processes so as to minimize public involvement.[35]

To compound the problem, industry representatives are also under no legal obligation to release information regarding the true costs of facilities in addition to health and safety risks.[36] In the pioneering days of nuclear power, French planners enjoyed the luxury of developing nuclear plans "without any public involvement."[37] Disregard for the need to inform the public on critical matters relating to nuclear power was even more pronounced in the former Soviet Union. When the Mayak Industrial Reprocessing Complex in the Southern Urals suffered a devastating accident, the Soviet government evacuated the 272,000 people living around the facility but told them that it was only an exercise. After the Chernobyl accident in 1986, the government did not begin evacuations until April 28 – two full days after the accident – because plant operators had delayed reporting the accident to Moscow out of fear that it would spoil the forthcoming May Day celebrations. Even after learning of the disaster, state officials planned on covering up the full extent of the damage, until a Swedish radiation monitoring station 800 miles northwest of Chernobyl reported abnormally high radiation levels.[38] Birth defects related to Chernobyl continue to this day, as figure 6.3 reveals. Radioactive pollution via both forest fires in the red zone and a legacy of birth defects have culminated in cases of "Chernobyl Heart" and deteriorating public health across Belarus and Ukraine.[39]

Nor are anti-democratic tendencies confined to the nuclear power industry. The indigenous people of Canada – the Dene, Cree, and Metis, inhabitants of North America for 12,000 years – have been repeatedly displaced by the expansion of oil tar sands production in Alberta.[40] In the United States, recent protests against the Keystone XL and Dakota Access pipelines highlighted the effect of energy infrastructure construction on the natural landscape and the sovereignty of Native American tribes.[41] The Dakota Access pipeline, for example, is partly built through the Standing Rock Sioux tribe's ancestral burial grounds and threatens water quality for many members of the tribal community. On multiple occasions, however, unarmed protesters have allegedly been faced with police and private security brutalities.[42] Figure 6.3 (a) also shows the toxic landscapes in Azerbaijan where life expectancy is almost half that of most

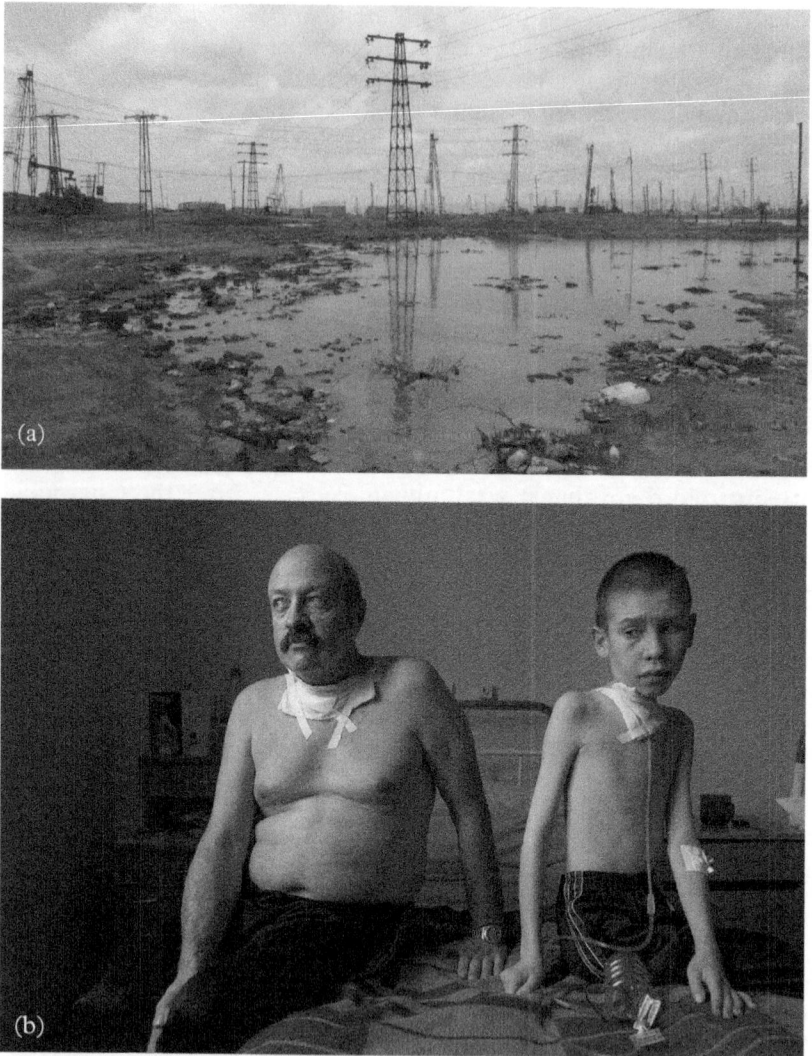

Figure 6.3 Legacies of energy injustice in Azerbaijan and Eastern Europe: (a) Average life expectancy for someone living near these oil fields near Baku, Azerbaijan, is less than 40 years compared to a global average above 70 years; (b) Oleg Shapiro (aged 54) and Dima Bognanovich (aged 13) receive treatment at the Thyroid Center in Minsk

Source: Authors.

industrialized countries. And in Oaxaca, Mexico, indigenous land-owners will receive a meager 1.5 percent of the gross income from energy production from a series of wind farms under construction, with the remainder going to companies such as Cemex, Gamesa, and Iberdrola, even though some of the land utilized, such as Juchitan, was appropriated from communal property without consent.[43]

Human rights abuses and discrimination

Large, international, conventional energy companies – particularly oil and gas suppliers – have employed private security firms to protect their operations and suppress dissent. In Indonesia, Myanmar, Nigeria, and Peru, some firms selling oil and gas to the US market have denied free speech, employed torture, supported slavery and forced labor, sanctioned extrajudicial killings, and ordered executions.[44] Shell allegedly gave guns to Nigerian security forces, and Chevron provided aid, helicopters, and pilots to an armed group that then gunned down nonviolent protestors on an oil drilling platform. BP, ExxonMobil, ConocoPhillips, and Shell continue to provide daily "security briefings" for mercenaries, and supply vehicles, arms, food, and medicine to soldiers and police.[45] In Afghanistan, the Talisman Oil Company has also been directly involved in the civil war, with evidence of the deployment of mujahedeen and child combatants to provide security around oil and gas blocks.[46] Shell was also accused of providing the Nigerian army with vehicles, patrol boats, and ammunition, as well as helping plan raids and military campaigns against villages.[47]

Fossil fuels are not the only energy systems that may contribute to the degradation of human rights. Low-carbon transitions in mobility and electricity are effectively implicated in toxic pollution, biodiversity loss, exacerbation of gender inequality, exploitation of child labor, and the subjugation of ethnic minorities. In Ghana, for example, the waste streams from solar panels, electric vehicles, and energy-saving digital devices from around the world end up poisoning communities residing near electronic waste scrapyards; such waste also contributes to the discrimination of migrants and ethnic groups from Ghana's north.[48] In the Democratic Republic of the Congo, mining for cobalt and copper, also needed for batteries, electric vehicles, and the wiring and components used in all forms of renewable energy, contribute to child labor as well as torture, violent beatings, and death.[49]

Involuntary displacement

A fourth sobering manner in which energy systems harm justice is by involuntarily displacing local populations. For example, millions of individuals are involuntarily resettled every year due to energy projects. The largest two sources are hydroelectric dams and mining. About 4 million people are currently displaced by activities relating to hydroelectricity construction or operation annually, and 80 million have been displaced in the past 50 years from the construction of 300 large dams.[50]

This relocation, however, often proceeds without true and meaningful consultation and consent. In Sarawak, the largest state of Malaysia, the Sarawak Corridor of Renewable Energy (SCORE) are building no less than 12 hydroelectric dams connected to industrial facilities along the coast of Borneo.[51] Rather than provide electricity to the local population, the dams will instead supply all of their electricity to factories and smelters. The environmental impact assessments associated with the first of these dams, the Bakun Hydroelectric Project, were conducted only after construction commenced, and have been attacked for vastly underestimating the facility's negative effect on water quality, ecosystems, and communities.[52]

Halfway across the world, Turkey, Syria, and Iran have all heavily dammed the headwaters that flow into the Tigris and Euphrates Rivers for electricity and irrigation. The combined result is much less water for Iraq, leaving the Shatt al Arab river without enough supply for livestock, crops, and drinking. Tens of thousands of Iraqi farmers have already had to abandon their fields and homes for lack of water.[53]

In Indonesia, state-owned coal-mining enterprises, responsible for most production, are poorly regulated and have extensively displaced communities without consultation or consent. Part of this involuntary resettlement has to do with Article 33 of Indonesia's constitution, which states that the natural resources of the country are to be exploited under state control – with no need for informed consent – for the maximum benefit of the people of Indonesia. The Ombilin coal mine, with reserves of 109 million tons, has disregarded local land rights entirely and improperly managed acid rock drainage and sediment ponds to the degree that 60 square kilometers of land had to be abandoned by indigenous communities.[54]

More broadly, one wide-ranging international survey estimated that at least 2.6 million people have been displaced due to mining in India from 1950 to 2009, and individual mines in Brazil, Ghana, Indonesia, and South Africa have involuntarily displaced between 15,000 and 37,000 people from their homes.[55] That study noted two troubling conclusions. First, the impact of such displacement extends beyond loss of land, which represented only 10–20 percent of impoverishment impacts, to include joblessness, homelessness, marginalization, food insecurity, increased health risks, social disarticulation, and the loss of civil and human rights. Second, it found that compensation for resettlement rarely occurred, and that when it did, such compensation was insufficient to restore, let alone improve, quality of life for the displaced.

Environmental racism

Many forms of energy production promote "environmental racism." Environmental racism "refers to any policy, practice, or directive that differentially affects or disadvantages (whether intended or unintended) individuals, groups, or communities based on race or color."[56] Dirty infrastructure can sometimes create "national sacrifice zones" that condemn poorer communities and people of color to suffer disproportionately.[57] Most of those working in energy-related jobs with greater occupational hazards (such as coal mines or refineries) tend to be near the poverty line; poorer families have less capital to invest in energy efficiency and thus live in homes that consume more electricity; and lower income families live in neighborhoods in closer proximity to conventional power plants, high voltage transmission lines, nuclear reactors, municipal landfills, trash incinerators, pipelines, abandoned toxic dumps, and nuclear waste repositories (and thus are more exposed to the life-endangering pollution that they bring).[58]

In Eastern Europe, the Roma have been displaced from so many countries and cities that they are forced to reside in settlements akin to "environmental time bombs." Roma communities in the Czech Republic and Slovakia, for example, reside in flats located above abandoned mines, which are prone to flooding, and they are susceptible to breathing methane gas. Others live in abandoned factory sites surrounded by mining waste where children are fully exposed to toxins and suffer long-term effects on health.[59] The situation with landfills in Scotland and nuclear waste storage facilities in Taiwan have also followed a similar trend, with studies confirming that the poor suffer a "triple jeopardy" of being most exposed to higher levels of pollution, being more vulnerable and more likely to suffer health impacts, and being least responsible for generating air quality problems in the first place.[60]

One study concluded that current nuclear power and waste sites, as well as dumps, mines, and other energy facilities, express characteristics that make them "peripheral communities" set apart from vibrant (and often wealthier) urban communities.[61] These peripheral communities tend to be:

- *remote,* either geographically separated from population centers or relatively inaccessible;
- *economically marginal,* with most being homogeneous in terms of social and demographic background and dependent on the nuclear industry as a dominant employer;
- *politically powerless,* with most key political decisions being made elsewhere, often in metropolitan centers;
- *culturally defensive,* with residents expressing ambivalent or ambiguous attitudes toward nuclear energy combined with feelings of isolation and a fatalistic acceptance of nuclear activities;
- *environmentally degraded,* meaning they tend to occupy previously polluted land or are close to places where radioactive risks are already present.

In essence, nuclear facilities will invariably migrate to communities that lack the political, social, and economic strength to oppose them, especially indigenous peoples and tribes, often at the extreme social and geographical periphery of society.[62] This process has been documented not only in Europe, but also in newer nuclear power regimes in Japan and South Korea.[63]

Even in the United States, injustices are apparent. More than two-thirds of all African Americans live within 30 miles of a coal-fired power plant. They are rushed to the emergency room for asthma attacks at more than four times the national average, and have children three times as likely to be hospitalized for treatment of asthma.[64] About half of African American children have unacceptable levels of mercury and lead in the bloodstream compared to 16 percent of the general population, and nationwide studies demonstrate that the air in communities of color contain higher levels of PM, carbon monoxide, ozone, and SO_2.[65]

Again, we do not wish to vilify only fossil fuels and nuclear power in these circumstances. In South Africa, the bidding and licensing process for renewables such as wind and solar tended to marginalize local communities.[66] In Gujarat, India, land acquisition processes for solar energy power plants (called "parks") have exacerbated the precariousness of vulnerable communities, who were left to suffer loss of livelihoods due to the enclosure of common land and extra-legal mechanisms through which land acquisitions for the project have reportedly taken place.[67] There, solar energy has only created "a regime of accumulation," whereby "low-carbon coalitions of interests can maximize their gains by dispossessing vulnerable social groups of their life-sustaining assets." Biofuel cultivation for private firms has also been prone to grabbing land from local communities, farmers, or pastoralists across Ghana, Kenya, Madagascar, Senegal, South Sudan, Tanzania, and Uganda.[68]

Activists, social movements, and fossil fuels

In recent years, there has been a surge worldwide in initiatives, policies, and social campaigns whose primary focus is the link between fossil fuels and climate change, and that start from the idea that fossil fuels are inherently and morally "wrong," precisely because of their detrimental impact on the environment and climate change. Evidently, leaving large proportions of fossil fuels in the ground also raises important ethical questions about the geographical distribution of this "unburnable carbon" and it raises questions over what constitutes a "just" transition. This section will discuss the growing anti-fossil fuel activism, the ethics of unburnable carbon, and the "just transition" discussion.

"Keep it in the ground" movement

There is a global surge in anti-fossil fuel activism. Campaigners and grass-roots activists seek to promote new moral norms that limit and ultimately prohibit fossil fuel-related actions, processes, or products for violating basic ethics.[69] These campaigns, and the norms they promote, differ from earlier approaches to climate and energy campaigning in at least four ways. First, they address fossil fuels directly. The fossil fuel industry, in particular, is considered the "enemy" for today's activists.[70] Second, new activism formulates potential policy actions across the *entire* fossil fuel supply chain. That is, practices of fossil fuel financing, extraction, and processing ought to be problematized just as much as patterns of fossil fuel consumption. Third, unlike policies that prioritize an economic approach to curbing greenhouse gas emissions and fossil fuel demand, activists essentially challenge the "normality" of fossil fuels in our society through a legitimacy-based approach. Campaigns in essence seek to remove the industry's "social license to operate." Fourth, these campaigns and emerging anti-fossil fuel norms differentiate from other ideational phenomena in the climate regime such as "sustainable development" and "common but differentiated responsibility" precisely because of their explicit focus on fossil fuel prohibition and not just on reduction.

One example of a normative campaign is the transnational *fossil fuel divestment campaign*, which urges investors to withdraw their financial investments in fossil fuel bonds, stocks, and other assets, and to then reinvest them into green alternatives. Around the world, multilateral development banks, national development finance institutions, and export credit agencies are "greening" their development finance portfolios and shunning fossil fuel investments, especially in coal.[71] Some campaigners think that an active shareholder approach is a better strategy than to divest, as shareholders can hold the corporate management of fossil fuel firms accountable, and they can push for specific climate targets for these companies to set themselves, including, for example, CO_2 reductions or developing corporate scenarios aligned with a Paris Agreement scenario.[72] Some have even used the legal terminology of "liability" and argued that fossil fuel firms ought to already be putting compensation into a fund for future payout to communities negatively affected by the consequences of climate change.[73]

There are also policy initiatives and campaigns that *single out extraction and processing (or transportation) of fossil fuels*. For example, numerous EU countries and local governments have banned fracking in recent years; the US and China, the world's largest coal producers in the world, temporarily imposed moratoria on new coal mines in 2016; and the US and Canada, also in 2016, designated public-owned waters in the Arctic and certain areas in the Atlantic Ocean as indefinitely off limits to future oil and gas leasing, although this plan has come under scrutiny under the Trump administration.[74] Mid-stream fossil fuel activities are also subject to anti-fossil fuel campaigns. Protests in the US against the Keystone XL and Dakota Access pipelines are among the most visible examples of normative campaigns targeting fossil fuel transportation and processing. Around the world, anti-fossil fuel campaigners have also targeted coal ports (see, e.g., Code Rood that blocked the coal port of Amsterdam, or protests against expansion of Abbott Point coal terminal in Queensland, Australia).

Litigation is also an increasingly visible way of compelling governments and companies to take climate change action. High-profile cases such as Urgenda in the Netherlands – where a non-governmental organization (NGO) won a legal case to compel the Dutch state to take more effective action to address climate change – have spiked activist interest in climate litigation. Indeed, judiciaries and the litigants that prompt them to action can indeed play important governance and normative roles. The adjudication of climate change has the potential to shape norms and beliefs in the broader population regarding the salience of climate change and the responsibility of different actors to act, including states and the fossil fuel industry itself.[75] Examples include suing large fossil fuel companies for their decades-long active role in covering up evidence on their climate change impacts, or their failure to accurately disclose the risk of climate change (legislation) to their shareholders.[76]

Campaigns related to fossil fuel consumption seek to phase out demand for oil, coal, and gas. For example, in recent years, an increasing number of countries have set deadlines to phase out the sale of internal combustion engine (ICE) vehicles.[77] By the end of 2018, at least 16 countries had taken varying types of action to phase out ICE vehicles and increase the number of electric vehicles.[78] Other initiatives include protests against construction of new fossil fuel power plants, or campaigns to rapidly phase out existing ones. One such example is the international Powering Past Coal Alliance, which seeks to phase out the world's coal fleet by 2050 at the latest.

A last, recent campaign, has received considerable media and political attention: *the transnational "Climate Strike" movement,* with Swedish 16-year-old Greta Thunberg as its leading figure. This is a movement of school students who are demonstrating to demand stringent action to prevent further global warming and climate change.[79] Their argument is firmly embedded within issues of "intergenerational justice" given that, through their inaction, current politicians, policy-makers, business leaders, and other relevant stakeholders are failing to ensure a stable climate for future generations to enjoy a good life undisturbed by the damage our energy systems inflict on the world today. Other new social movements, such as Extinction Rebellion and the Sunrise Movement, are turning to more radical actions to support the same cause.

It is clear that there is an emerging set of morally inspired climate campaigns that single out fossil fuels, and in particular the fossil fuel industry, as key to rapid climate change and global warming. Although they differ in how "radically" they are organized and their objectives and arguments are formulated, they all recognize that the necessary energy transition away from a fossil fuel-heavy economy toward a low-carbon society also entails finding solutions to key questions about justice and equity.

Ethics of "unburnable carbon"

Evidently, leaving large proportions of fossil fuels in the ground also raises important ethical questions about the geographical distribution of this "unburnable carbon." Is there such a thing as the "right to extract" in a world where the objective is to limit global warming to "well below 2°C by the end of this century"? In the run-up to the negotiations at COP21 in 2015, for example, India's chief economic advisor decried the West's "carbon imperialism" and emphasized his country's right to choose a development path based on cheap and abundantly available coal.[80] Indeed, countries cannot be expected to curb their own extraction if it is not considered part of an effort that is fairly distributed over all fossil fuel producers (and consumers). A justice perspective on the allocation of fossil fuel extraction differs substantially from that of a conventional economic one, where "market forces" dictate where and which fossil fuels can be recovered at the lowest marginal production cost.[81] One set of questions where theories of justice can be relevant pertains to which assets can be extracted and by whom, while another set of questions refers to who should bear the costs of keeping fossil fuels in the ground, and who should enjoy the benefits of extracting the remaining permissible fossil fuels.

Regarding the first set of questions, it can be argued that gradually leaving fossil fuels in the ground ought to be slower in those countries where the rapid transition away from extraction would cause the greatest socio-economic disruption, that have the least viable alternatives for meeting their developmental needs (i.e. fossil fuels have historically provided a path to cheap and secure energy for development), and that have the least capacity to transition to these alternatives.

On the second set of questions, the costs of leaving fossil fuels in the ground can be distributed in accordance with two principles of equity. First, the principle of "historical responsibility" suggests that the greater the historical extraction, the greater the contribution to bearing the costs of keeping fossil fuels in the ground ought to be. This is also akin to a classic idea in environmental law of "the polluter pays" principle. Second, the principle of "capabilities" suggests greater capabilities imply a greater contribution to the costs. Capabilities can be understood as societies' economic, physical, and institutional capacities to cope with their own transition, as well as to provide support to others in coping with their transition.[82]

Neglecting the ethical implications of curbing GHG emissions and leaving fossil fuels in the ground comes with serious risks. They risk raising the possibility of advancing pathways to global decarbonization that are widely considered inequitable and therefore illegitimate. Climate change is a problem of the global commons, and a holistic, effective response requires global cooperation that is sufficiently robust to ensure a major global transformation. Cooperation, of course, will be less likely if countries do not consider the proposed solutions and policies to be reasonably fair.

Making a "just transition"

Importantly, justice considerations are also key components of the governance of the energy transition. These considerations are increasingly referred to as matters of a "just transition." While the precise definition various by source, core elements tend to consistently appear:[83]

- investments in establishing low-emission and labor-intensive technologies and sectors;
- research and early assessment of the social and employment impacts of climate policies;
- social dialogue and democratic consultation with social partners and stakeholders;
- training and skills development for exposed workers;
- social protection alongside active labor market policies;
- local economic diversification plans.

A just transition could require that the state intervene more actively in the political economy to create jobs in "green" sectors, in part to compensate for now-abandoned fossil-fuel-based sectors, and that state and capital (and those more able to pay higher associated taxes, for example) absorb carbon capitalism's negative social externalities, and provide a welfare safety net and adequate compensation for people and communities that have been marginalized or negatively impacted by a low carbon energy transition.

This concept originated as a trade union demand that has now become a mainstream policy tool applied by international institutions and treaties, as well as (sub)national governments that design environmental and climate policies. It refers to how the future and livelihoods of workers and their communities in the transition to a low-carbon economy can be secured and it is based on social dialogue between workers and their unions, employers, government, and communities.[84] For example, after years of negotiation, in December 2018, the Spanish government agreed to a €250 million just transition deal with coal miner unions. The deal included, inter alia, the closure of unviable privately operated coal mines, early retirement schemes, redundancy payments, retraining, as well as restoration and environmental regeneration of former mining sites.[85]

Conclusion

To conclude, the evidence in this chapter strongly implies that we view the *moral and ethical dimensions* of energy and the energy transition alongside the usual technical, economic, political, or even cultural ones. Many justice theorists would question the underlying morality of forcing or coercing people to abandon their homes for energy projects, imposing the burden of pollution on the young and vulnerable, violating basic human rights, misappropriating energy funds, and creating a low-carbon energy system with unequal access. This demands that contemporary analysts, policymakers, and even consumers begin to reconsider their energy decisions as moral ones. It also explains likely renewed emphasis on "just transitions" as well as the necessity of "energy justice."

A second core finding is that because all of us participate in the global energy system to varying degrees, each of us contributes to some – perhaps even many – of the problems and examples of injustice highlighted throughout this chapter. To be sure, responsibility is differentiated and complex and often beyond the direct reach of the individual consumer. Nevertheless, the decisions made about which home appliance to purchase, how much electricity to consume, and which car to buy have very real moral and ethical implications. The fuel oil or natural gas needed to heat homes or cook meals results in young children living near refineries inhaling health-endangering particulate matter. Electricity generation, particularly from coal-fired power plants, is partially responsible for the impacts of climate change – the tidal inundations and storm surges flooding low-lying cities – as well as the forcible relocation of indigenous communities near a coal mine or hydroelectric dam. Automobiles contribute to acid rain bleaching streams, forests, and coral reefs, or to human rights abuses in oil-producing countries.

As such, linking justice and energy together emphasizes the necessity of an *inclusive and participatory* approach not only to energy planning and policymaking, but the research process itself. Such inclusivity can enhance the robustness of research by incorporating academic knowledge about social structures, systems of cultural meaning, and processes of change as well as non-academics' "on the ground" ways of knowing as an instrumental "cognitive resource."[86] A diversity of methods, forms of knowledge, and stakeholders are needed to encourage a breadth of perspectives, especially those from beyond the academy.

PART II

Governing the energy transition

7
Energy technologies
and innovation

As we have seen in the previous chapters, our current, fossil-fuel based energy system spawns a host of problems for geopolitical stability, economic development, social justice and, most notably, accelerated climate change. If we do not decarbonize every sector of the economy in the coming decades, temperatures will likely rise above 2°C or more, with devastating consequences for our societies. Simple tweaks to our energy system will not be enough. We have to make fundamental changes to the way in which we produce, consume, and distribute goods and services. That is, we have to change the way we live.

Although there is a widespread recognition that we need to make some sort of global energy transition, many questions remain unresolved: What do we understand by the concept of an "energy transition"? Which technological options exist to make such a transition? What are the drivers and obstacles in the process of technological innovation and diffusion? Are energy transitions necessarily protracted affairs, or can they be expedited? This chapter will discuss these important issues. True to our energy systems perspective outlined in the introduction, we will not confine ourselves to the physical hardware of energy infrastructure, but we also consider (non-technical) demand-side options, closely associated with our personal behavior and lifestyles.

The good news is that we do not have to wait for technological breakthroughs. The low-carbon economy can be largely built by further developing proven technologies that exist already today. Many of these technologies – like solar PV, wind, and LED lighting – have seen dramatic cost declines and have reached the point where they are cheaper than their fossil fuel counterparts.[1] Yet, the bitter reality is that these new renewable energy sources are not growing fast enough to bend the global emissions curve. After having briefly plateaued in the period 2014–16, energy-related CO_2 emissions rose again in 2017 and 2018.[2]

The nature of technological innovation

An energy transition can be defined as a change in the overall energy system, usually to a particular fuel source, technology, or prime mover (a device that converts energy into useful services, such as an automobile or television) but also to new markets, new energy services, and new regimes.[3] Some studies choose to focus only on the first of those dimensions – a change in fuels such as oil, coal, gas, and nuclear energy – causing some to critique that they narrowly frame transitions as a way of foreclosing future change[4] or of masking "the social and political dimensions of energy systems behind a false veneer of limited technological choices."[5] Others take a broader view that encompasses shifts in technology as well as the resulting "constellation of energy inputs and outputs involving suppliers, distributors, and end users along with institutions of regulation, conversion and trade."[6]

Any discussion about transitions, and the technologies currently available to mitigate carbon, or accelerate low-carbon energy transitions, must first make a distinction between commercially available (and truly viable) options, and those that are less certain, more distant, and on the so-called frontier. Ultimately, this results in four different types of energy systems:[7]

- **Typically available** describes the traditional systems already used around the world to provide energy services, many of them fossil-fueled.
- **Currently available best practice** represents the most advanced commercially available climate mitigation technologies that are cost-effective and widespread today.
- **State-of-the-art feasible technologies** are defined as the best-performing technologies being prototyped and demonstrated that are technically feasible but have not yet been proven and indeed may not yet be cost-competitive.
- **Frontier or breakthrough technologies** are those that could perhaps some day result in significant emissions reductions, but are not yet even being piloted or trialed.

"Suitable for Low-Carbon Transitions"

	Typical Cuttent Use	Today's Best Practices	Technically Feasible Technologies
Demand-side	T-12 Fluorescent Office Lighting	LED lighting	Solid State and Hybrid Solar Lighting
	24 mpg Passenger Vehicles with Internal Combustion Engines	80 mpg Hybrid Electric Vehicles (HEVs)	Plug-in Electric Vehicles
Energy supply-side	30–35% Efficient Pulverized Coal Plants	70% Efficient Integrated Gas CHP	Solid Oxide Fuel Cells
	Gasoline Fuel for Motor Vehicles	Gasoline Blended with Ethanol (E10, E85)	Cellulosic Ethanol Blends
Carbon Capture and Sequestration	Unmitigated CO_2 emissions from power plants	Post-Combustion Amine Scrubbing	Integrated Gasification Combined Cycle with Sequestration
	Deforestation	Advanced Forest Management	Plants Bioengineered to enhance sequestration

Frontier or Breakthrough Technologies		
Small modular reactors (SMRs)	Organic solar PV	Bio-Catalysis
Fusion	BECCs	Algal fuels
Microbial Genomics	DACCs	Condensed Matter Physics

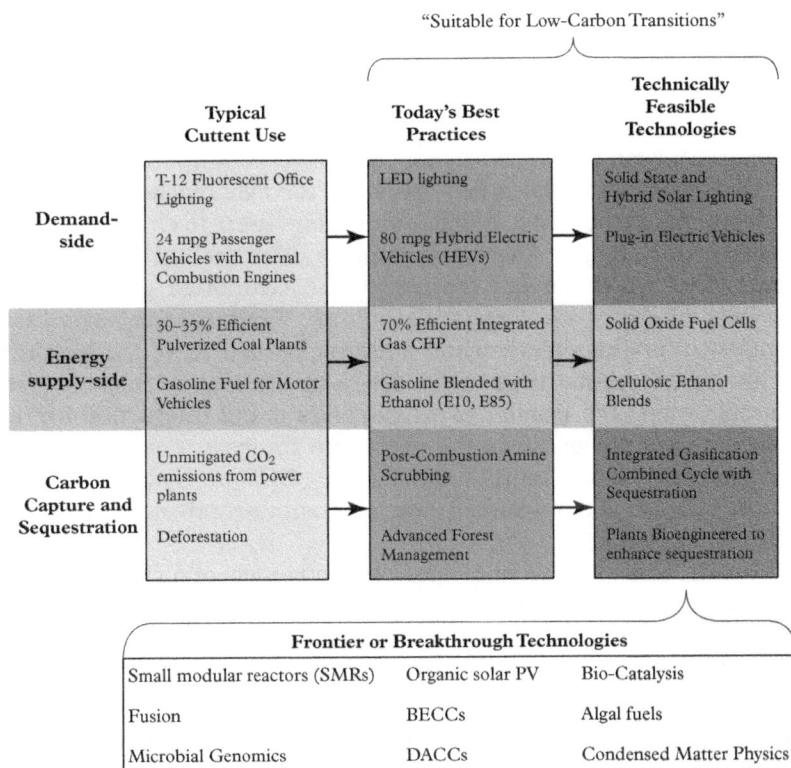

Figure 7.1 Currently available and state-of-the-art low-carbon energy systems

Source: Modified from Brown, M. A., & Sovacool, B. K. (2011) *Climate Change and Global Energy Security: Technology and Policy Options.* Cambridge, MA: MIT Press.

Based on this typology, this chapter assesses the two stages of technologies that are most important, i.e. those that are either currently deployable or nearly ready. The chapter divides such currently available and feasible technologies into three categories: demand-side options that involve energy end-use or practices; energy supply options such as renewables, hydrogen, and nuclear power; and carbon capture and storage options.[8] Figure 7.1 offers an overview of these technologies.

New energy systems or technologies can be further categorized according to whether they represent sequential innovations or disruptive innovations. Indeed, classic economic theory drawing from Schumpeter even supposed that the capitalist economy overall advances through a process of "creative destruction," where old, inefficient firms or technologies are replaced, or destroyed, by new emerging innovations.[9] A *disruptive innovation* represents a radical disjuncture that changes the very nature of the socio-technological regime. An example is the electric car, which does not need conventional fuel stations, has much less need for maintenance than cars with internal combustion engines, and allows users to charge the cars themselves, at home. Another example is decentralized and off-grid renewables, which disrupt the traditional, centralized grid and the business model of the large-scale utilities. *Sequential innovation*, by contrast, represents incremental adjustments that leave the existing technological regime unchanged. Deep-sea drilling or fracking, for example, shifts the frontier of oil exploration and production but it does not alter the basics of the petroleum business. For some technologies, like carbon capture and storage (CCS) or hydrogen, there is much debate about whether they represent disruptive or sequential types of innovation.

Usually, when new technologies are first launched, niche markets – early adopters – will begin to consume or use the technology. As awareness of the technology increases and uncertainty decreases, adoption begins to pick up until it reaches a plateau; this model is generally referred to as simply "the S-curve." Historically, the diffusion of technologies like the telephone, microwave ovens, and air-conditioning has broadly followed this pattern.[10] However, progress along this hypothetical curve is not guaranteed.[11]

A useful way to think about transitions in the energy system is to look at the three main uses in total final energy consumption: power (20 percent), transport (32 percent), and heating and cooling (48 percent).[12] Each of these three end-uses is dominated by a different type of fuel, and poses distinct challenges for decarbonization:

- About one-fifth of the energy that people use comes through an electric wire, that is, in the form of *electricity*. Most of that electricity is generated through the combustion of coal (38.3 percent), followed by natural gas (23.1 percent).[13] There are three main ways to decarbonize the power sector: through expansion of renewables, nuclear power, or CCS.
- *Transportation* today relies almost exclusively on oil (92 percent), refined into gasoline, diesel fuel, and jet fuel.[14] There are two main approaches to replace that oil in light-duty vehicles: one is to switch to biofuels, the other is to switch to electric vehicles.

- *Heating and cooling* is the other big end-use sector, and one that is often overlooked. It is currently dominated by natural gas and oil. It is worth separating heating into two basic categories: space heating (mostly for residential and commercial buildings) and process heating (mostly for industry). For the former, a combination of passive building design, electric pumps and district heating offers a set of workable and affordable solutions.[15] Many of these solutions also work for the latter, and they can be complemented with biomass, biogas, and electricity for process heat applications requiring high temperatures (generally, above 400°C).[16]

Seen from this perspective, the challenge of decarbonization thus requires first and foremost to phase out oil in the transport sector, phase out coal and gas in the power sector, and phase out gas and oil in the heating sector. A large part of the solution will probably consist of "sector coupling" – that is, interconnecting the power, transport, and heating sectors. The electrification of passenger vehicles, for instance, not only offers a way to decarbonize transportation, but also offers useful services to a grid that is increasingly powered by intermittent renewables (as "batteries-on-wheels"). Renewable electricity can also be used to produce hydrogen or synthetic gas ("power-to-X"), which can in turn be used to power vehicles or it could be turned back into electricity or heat during times of little sun and wind.

In the next subsections, we review some of the technological options to make an energy transition. The first place where the transformation is occurring is on the "demand side" of the energy system, the second is on the "supply side."

Demand-side options

The first category of options all relate to the "demand side," or "energy use." This is often described as "energy efficiency" and "managing energy demand." Energy efficiency does not necessarily mean "doing less" or "suffering without," but instead what physicist Amory Lovins calls "doing more with less through smarter technologies."[17] It does not mean cold showers and warm beer. Rather, energy efficiency refers to getting more energy services per unit of energy consumed – with light bulbs that need less power, weather stripping around doors and windows, hybrid electric vehicles instead of gas-guzzlers, properly inflated automobile tires, more

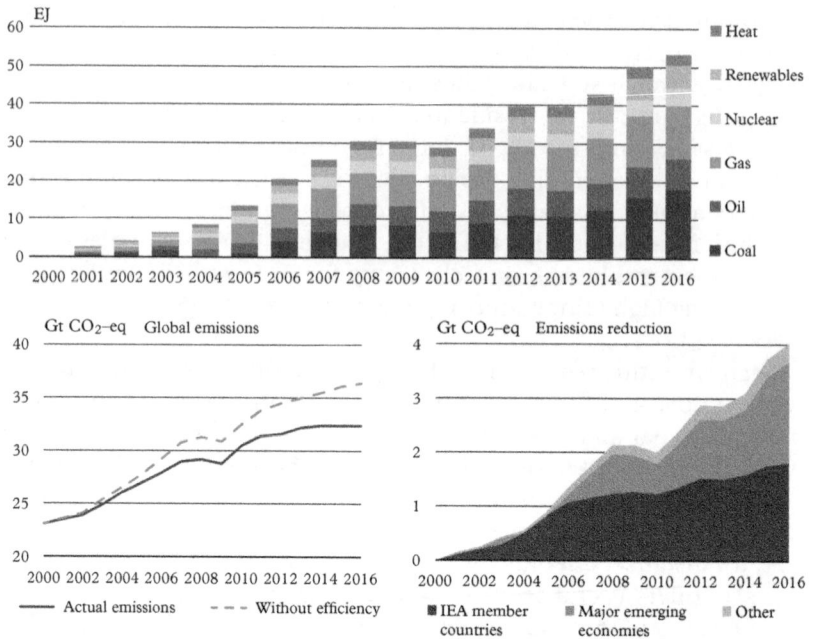

Figure 7.2   Avoided primary energy demand and carbon emissions from energy efficiency improvements, 2000–2016

efficient industrial motors, and the recovery of waste heat in industrial plants.

In characterizing the energy mix, both the IEA and the EIA use categories of fuels such as coal, oil, natural gas, nuclear, and renewable resources. The omission of energy efficiency from this mix reinforces the perception that a megawatt saved (i.e., a "negawatt") is not as valuable as a megawatt generated. Yet numerous studies indicate just the opposite. As figure 7.2 reveals, the avoided energy demand from energy efficiency improvements in the period 2000–16 surpassed 50 EJ, equivalent to adding another European Union to the global energy market, and these savings corresponding to an avoided 4 billion tons of CO_2 equivalent.[18]

Specific energy-efficient technologies can be loosely divided into two subcategories most relevant here: those used in transportation, on the one hand, and buildings, on the other.

High-efficiency transportation

On a global scale, the transportation of people and goods accounts for approximately one-quarter of the world's energy consumption and one-quarter of its energy-related CO_2 emissions.[19] Over the next few decades the transportation sector is expected to be one of the fastest-growing sources of GHG emissions. Much of the projected increase is attributed to the rapidly growing demand for petroleum-based transportation fuels in non-OECD economies (i.e., the developing world), which are forecast to increase more than 2 percent per year; as compared with the OECD countries, which are forecast to increase at less than 1 percent per year.[20]

Many technically feasible technologies and practices can reduce the consumption of petroleum transport fuels. These include light-weight materials, cylinder deactivation, electric-fuel engine hybrids ("hybrid electric" vehicles and "plug-in hybrids"), clean diesel engines, and the use of hydrogenated low-sulfur gasoline. In aviation – the next largest category of transportation energy demand after highway use – GHG emissions could be lowered through improved engine designs, fuel blends, and air traffic management systems. Finally, reductions could result from modal shifts (e.g., from highway modes to rail, facilitated by improved intermodal connections), higher load factors, more intelligent transportation systems, better freight hauling efficiencies, reduced idling by heavy-duty vehicles, and more sustainable land-use configurations with mixed uses and denser developments.

Several vehicle designs have either already begun to sell into mass markets or have near-term potential to penetrate the market. The sale of gasoline-electric hybrids, for example, has grown significantly in recent years, while plug-in hybrid electric vehicles (PHEVs) are capable of even greater fuel economy and gasoline displacement with electricity. For example, in 2018, the global electric vehicle fleet exceeded 5.1 million passenger cars, up 2 million from 2017 and almost doubling the number of new electric car sales, with the top three largest markets for EVs being (in order) China, the United States, and Europe.[21] Lithium-ion batteries provide electric power for existing and next generation hybrids that continue to be developed. Indeed, ERG, a battery manufacturer, estimates that due to the Paris Agreement, a 50-fold increase in electric vehicle adoption needs to occur between 2016 and 2030, reaching 100 million electric vehicles by that year. Thus, the cobalt demand in electric vehicle batteries is expected to grow by 200 percent between now and 2020, and again by 500 percent by 2025, when the battery market is expected to be worth \$100 billion.[22]

Flex-fuel vehicles, which can utilize ethanol or gasoline, are increasingly available today. The technology for other alternative fuel vehicles, such as biodiesel, natural gas, and propane, currently exists for commercial use, as well. In addition, prototype hydrogen fuel cell and hydrogen internal combustion engine vehicles, as well as hydrogen-refueling infrastructure, have been developed and demonstrated.[23]

Lastly, more intelligent transportation systems can reduce traffic and increase fuel efficiency, such as high-speed and automated toll collection, adaptive signal controls, incident management systems, and travel information systems. Advanced screening technologies can make inspection of commercial vehicles more efficient and save fuel. Transit-oriented development and mixed land use urban designs also hold significant potential for reducing the miles traveled by vehicles.

High-performance buildings

The built environment – consisting of residential, commercial, and institutional structures – accounts for about one-third of primary global energy demand[24] and is the source of 35 percent of global energy-related CO_2 emissions.[25] Over the long term, buildings are expected to continue to be a significant component of energy use and emissions, driven in large part by the continuing trends of urbanization, population, GDP growth, and the longevity of building stocks. A growing body of evidence suggests that improving the energy efficiency of the existing building stock and new construction is a low-cost approach to mitigating GHG emissions.[26]

Fortunately, lighting, office equipment, appliances, and even components of the building structure such as windows, doors, and insulation are typically replaced and upgraded on a frequent basis. Therefore, the short-term potential for improving the energy efficiency of the existing building stock is high. In both China and the US, the single largest user of energy in residential buildings is space heating, followed by water heating and appliances in China, and by other uses (primarily electric appliances) and water heating in the US.[27]

While the built environment is a complex mix of heterogeneous building types and functional uses, all have common features, each of which may benefit from energy-saving technological advances, thereby reducing CO_2 emissions from buildings. On the appliance and equipment side, there are numerous GHG-saving opportunities:

- Heating and cooling advances include air and ground source electric heat pumps, gas-fired absorption heat pumps, centrifugal chillers, desiccant air pre-conditioners, and integrated heat pump systems that provide cooling, heating, hot water, and dehumidification, merged into single pieces of equipment to lower cost and increase efficiency.
- Instantaneous water heaters and heat pump water heaters are also commercially available but not widely used. In addition to being highly efficient, heat pump water heaters are particularly attractive in hot and humid environments because they also dehumidify the surrounding air.
- Solar systems for heating air and water have been commercially available for decades. While solar space heating has seen limited usage to date, solar water heating has taken hold in many countries with strong solar radiation including Spain (which mandates solar water heating nation-wide) and China (where one of its most populous provinces, Jiangsu, requires solar water heating in all new low-rise residential buildings).[28] One of the most popular uses in the US is for pool heating, where solar heating accounts for a majority of the heated pools sold in recent years.[29]
- Solid state lighting (including light-emitting diodes or LEDs) is an important emerging technology with significant energy-savings potential. It produces light by passing electricity through a semiconductor material, which produces very little heat and therefore is a highly efficient form of light.[30]
- Various electronic lighting controls, including dimmers, motion sensors, occupancy sensors, photo sensors, and timers are also commercially available but they have not cornered a dominant share of the marketplace.
- Combined heat and power (CHP) systems (sometimes referred to as "cogeneration") hold great promise. It is a form of distributed generation that requires much less fuel to achieve the same energy output as separate heat and power systems.
- Improved shell designs and insulating systems can reduce energy use significantly, just as improved training of construction professionals can enable builders to achieve the promise of these gains.
- The addition of energy storage systems – from more conventional batteries, flywheels, pumped storage, and compressed air energy storage to more advanced superconducting magnetic energy

storage, supercapacitors, hydrogen or molten salt – can offer a variety of additional benefits to high-performance buildings. These include greater reliability via uninterrupted power supply, the ability to sell power back to the grid at peak times, the option of selling grid services alongside electricity, and bulk power management.[31]

Through technology advances such as these, buildings worldwide can become part of the climate solution. This is the vision of the zero-energy home, factory, city, or even society.

Low-carbon energy supply options

Transitioning away from increasingly scarce, high-emissions fossil fuels to low-carbon energy supplies is one of other twin challenges facing energy and climate planners. Prominent among the energy supply options with inherently low life-cycle CO_2 emissions is a suite of renewable technologies. To the extent these technologies emit GHGs, the emissions generally occur during manufacturing and deployment and not during the combustion of fuels.

The inherently low-carbon nature of these technologies comes from the fact that most renewable technologies are powered by the sun:

- Plants and algae require sunlight for photosynthesis before they can be converted to biofuels or biopower.
- Hydropower capitalizes on rain and snowfall from water evaporation and transpiration.
- Wind generates electricity directly by turning a turbine or indirectly in the form of ocean waves, but the wind itself is driven by the sun.
- Ocean thermal energy conversion uses the temperature differential between surface water warmed by the sun and cold deep water to drive a turbine and make electricity.

Tidal and geothermal energy are the only renewable energy resources that are not a direct result of solar energy. Tides go up and down due to the gravitational attraction between the oceans and the moon. The heat trapped in the earth itself is due to both leftover heat from formation of the planet and the radioactive decay of elements within the crust, such as uranium and thorium. Energy supply options with relatively small carbon footprints also include hydrogen for power production and fuels (if the hydrogen is produced sustainably), and nuclear power (depending on the uranium production and spent fuel management).[32]

Renewable electricity

Renewable energy deployment is expanding at double-digit rates across the globe.[33] Although they are starting from a small base, renewables are the fastest-growing energy source worldwide.[34] Much of the growth is in hydropower, solar PV, wind power, and biomass. Hydropower, onshore wind turbines, solar PV, and electricity from landfill gas, have generally achieved "grid parity"– that is, they have reached the point where they can generate electricity at, or below, the cost of traditional grid power. Other renewable technologies are not cost-competitive under current pricing regimes. As a result, government policies and incentives typically are the primary drivers for the construction of renewable generation facilities.[35]

Renewable technologies for generating electricity are generally modular and can be used to help meet the energy needs of a standalone application such as a building, an industrial plant or community, or the larger needs of a regional or national electrical grid network. Because of this flexibility, interconnection standards (for connecting individual renewable technologies with individual loads or buildings and with the electric grid) are critical to the safe, reliable, and efficient performance of renewable systems.

In addition to the flexibility of hybrid fossil/renewable systems, the diversity of renewable energy sources offers a broad array of technology choices that can reduce CO_2 emissions. The generation of electricity from solar, wind, geothermal, or hydropower sources contributes essentially no CO_2 or other GHGs directly to the atmosphere, and renewable resources generally emit fewer GHGs than conventional fuels. Increasing the contribution of renewables to any nation's energy portfolio will directly lower GHG emissions in proportion to the amount of carbon-emitting energy sources displaced.

The technologies in the suite of renewable options are in various stages of market readiness. Within solar, wind, geothermal, ocean, biomass, and hydropower, each resource includes mature technologies that either have already been commercialized or are suitable for near-term commercialization. Advocates of renewables suggest that claims of insufficient renewable energy are invalid. Consider, they say, what we know about the potential of some of the dominant renewables.

The sun beams down more energy to the Earth's surface in an hour than the world uses in a year.[36] The World Ocean Report, which examines the sustainable use of energy resources provided by the sea, has estimated that offshore wind power alone could supply approximately 5,000 TWh of electricity each year. The same report estimates that a further 1,700 TWh of electricity could be produced from readily exploitable wave energy. This total represents over 40 percent of current annual global electricity consumption.[37] The IPCC contends that nearly 80 percent of the world's energy needs could be met by renewable energy by 2050, if the right public policies were put in place to support such an initiative.[38] IRENA notes more recently (in 2018) that the deployment of renewables will likely increase to make up two-thirds of energy consumption and 85 percent of power generation in 2050.[39]

Many of the renewable technologies that possess the greatest utility-scale potential do not produce electricity all the time. Sometimes wind blows; at other times, it does not. Sometimes the sun shines; but not always. Wind power, solar power, wave power, tidal power, hydropower, and even biomass power (due to seasonal changes) all suffer from intermittent or stochastic power profiles.[40] This makes these forms of energy somewhat of a burden for utilities to manage. The task of providing power to energy users is hard enough given variable demand patterns. When fluctuating supply patterns must also be added to the planning challenge, the task of balancing electricity production with electricity demand can be expensive and can potentially undermine the quality of electricity services provided. The stochastic profile of renewable energy flows is not necessarily an unavoidable technological problem; however, it is only a technological problem due to the nature of the systems that are currently in place for electricity load balancing.

However, coal plants suffer unexpected outages more commonly than one might think; nuclear plants need to go offline for more than a month when they refuel; natural gas plants often run in peaking mode which means they operate sparsely. Indeed, the US Department of Energy reports that the average capacity factor for all power plants across the nation, a reflection of technical reliability as well as dispatching practice, is roughly 55 percent. That is, over a long period of time, an average power plant actually contributes to the electricity grid only 55 percent of its theoretical maximum output.[41]

Moreover, over the past decade, concerns over how to manage the stochastic flows of renewable energy generation have weakened significantly due to improved energy portfolio management strategies, more effective plant siting strategies, and technological advances within many of the renewable energy platforms. New approaches to load-balancing strategies are appearing in engineering journals and new strategies are being advanced to reduce stochastic variation through strategic siting of renewable energy plants and technical portfolio combinations such as twinning solar and wind. Overall, improvements to renewable energy technologies and strategies for better employing storage technologies are beginning to make the stochastic flows of renewable energy generation far less of a concern.

Renewable transport

Renewable forms of energy can be harnessed not only to generate power, but for mobility as well. The two most dominant biofuels are ethanol and biodiesel.[42] Ethanol is derived by fermenting grains, cereals, sugar crops, and other starches, predominantly corn and sugarcane. Crushing and soaking practices remove the sugar from these crops, which is then fermented in alcohol using yeasts. Biodiesel can be made from vegetable oils, animal fats, or microalgae oils. Feedstocks for biodiesel, by contrast, tend to be oil-rich crops such as soybean, jatropha, palm oil, rapeseed (canola), and sunflower seeds, or waste from cooking oil and animal fats. Most biodiesel plants employ transesterification, a process of mixing an oil feedstock with an alcohol via a catalyst. Biodiesel in particular can be blended with conventional diesel fuel or used as a neat fuel (100 percent biodiesel).[43] Biodiesel emits 64 percent fewer GHGs than diesel gasoline in terms of pounds/gallon produced from the production and use of fuel.[44] Though definitions about "first" through "fourth" generations persist,[45] table 7.1 summarizes a commonly accepted distinction between "conventional" biofuels and "advanced" biofuels.

The pros and cons of biofuels are perhaps more starkly debated than the other options in this chapter. Critics point out that first-generation biofuels can replace only a small sliver of global petroleum supply, meaning they cannot truly eliminate dependence on fossil fuels.[46] Even then, expanding production of biofuels, usually through a few select agricultural monocultures dependent on pesticides and fertilizers, poses adverse effects on the global climate, on land and water resources, and even on food prices and global food security. When peatlands or rainforests are cleared to make way for biofuel

Table 7.1 Attributes of conventional and advanced biofuels

	Stage	Nomenclature	Fuel types	Feedstocks
Conventional	Commercially available or nearly available	"First" generation	Sugar- and starch-based ethanol, oil-crop based biodiesel and straight vegetable oil, as well as biogas derived through anaerobic digestion	Sugarcane and sugar beet, starch-bearing grains like corn and wheat, oil crops like rape (canola), soybean and oil palm, and in some cases animal fats and used cooking oils
Advanced	Basic and applied R&D or demonstration only	"Second," "third," or even "fourth" generation	Cellulosic ethanol, algal fuels (diesel from microalgae), biomass-to-liquids (BtL)–diesel, furanics, gasification	Genetically engineered crops such as switchgrass, miscanthus, or jatropha. Hydrotreated vegetable oil (HVO), which is based on animal fat and plant oil, as well as biofuels based on lignocellulosic biomass, such as cellulosic-ethanol, BtL–diesel and bio-synthetic gas (bio–SG). Also included are algae-based biofuels and the conversion of sugar into diesel-type biofuels using biological or chemical catalysts as well as gasification

Source: Modified from Sovacool, B. K., Brown, M. A., & Valentine, S. V. (2016) *Fact and Fiction in Global Energy Policy: Fifteen Contentious Questions.* Baltimore, MD: Johns Hopkins University Press, p. 163.

plantations, the greenhouse gas emissions released are climatically catastrophic. Lastly, critics point out that such biofuels can have higher costs than other forms of carbon reduction and lead to harsh agricultural labor conditions.

Advocates of biofuels state equally forceful reasons to support their view. Done right, biofuels have the capacity to displace the use of fossil fuels, especially the importation of oil from unstable geopolitical regions. They release no fossilized carbon, can sink carbon through their roots and trees, and generate waste products that can be harnessed for things like electricity, meaning a single feedstock can produce multiple energy services. In some cases, they can produce organic fertilizer that displaces chemicals. Biofuel industries can also support and revitalize rural areas, enhance agricultural development, and reduce other forms of air pollution such as acid rain. As one peer-reviewed study chimed:

> Biofuels as an alternate source of energy have gained immense popularity globally, as they are perceived to be a clean alternative energy source which can be developed inexpensively, through improved technology. They have a further advantage over conventional fuels which require an extremely long process of restoration, in that they can be obtained in a matter of days, and in all three states of matter – solid, liquid and gas. Biofuels being plant-based do not emit sulfur or carbon dioxide and are thus a non-toxic, non-polluting, bio-degradable and environmentally friendly fuel.[47]

The next generation of advanced biofuels, such as cellulosic ethanol, can be made from compounds in green plants such as stalks, leaves, grasses, and even trees; specially engineered crops for biofuel manufacturing, thereby reducing GHG emissions significantly;[48] or algae and mold, capable of producing 10–30 times more oil per acre than crops currently in circulation.

Hydrogen

Hydrogen has the potential to be an attractive non-carbon energy carrier for both the transportation sector and stationary applications through the use of fuel cells. The use of hydrogen fuel cell vehicles can achieve large reductions in oil consumption and CO_2 emission, especially if they draw from solar or wind electricity, although several decades will be needed to convert this technical potential into achievable benefits.[49] As such, there is widespread international interest in advancing hydrogen technology to address the growing emissions, supply, and energy security concerns associated with conventional fuels. Governments and industry around the world are engaged in research and development programs to enable hydrogen technology to be a viable and cost-effective alternative. In the US, the strategy for deploying hydrogen-based technologies focuses primarily on the transportation sector, while Japan and other countries are focusing more on hydrogen used in stationary fuel cells for residential heating and cooling. The IEA notes that demand for hydrogen has expanded more than threefold since 1975, and that it currently accounts for 6 percent of natural gas consumption worldwide and 2 percent of coal production.[50]

Nuclear power

Nuclear power, more precisely defined as nuclear fission, is today a significant source of low GHG-emitting "baseload" electricity production, depending on the fuel cycles of particular plants. Nuclear power plants are particularly well suited for generating "baseload" power, because they are most economical when operated at or near full capacity and with little load-following variability.

Four major challenges to increased use of nuclear power remain, including cost, safety, waste management, and proliferation risk.[51]

- The cost of electricity from new nuclear plants is expected to be higher than the cost of new gas-fired and coal-fired power along with low-cost renewables in most countries. Nuclear power's position will improve if a price is placed on GHG emissions. It is also less costly in countries like South Korea that have been able to construct its last six nuclear units in only 53–58 months.[52]
- Despite the improved safety record of today's nuclear reactors, if the number of reactors is to grow twofold or more worldwide, defenses against terrorist attack as well as malfunction and human error need to be improved.[53]
- Despite the fact that both open and closed fuel cycles require the geologic disposal of some radioactive waste, little progress in high-level waste disposal has occurred.[54]
- The risk of the proliferation of weapons-grade radioactive materials, to possibly develop nuclear weapons, remains an issue worldwide, especially with commercial nuclear energy programs in Iran and North Korea.

Table 7.2 Commercially available and technically feasible CCS options

	Commercially available on the market today	Technically feasible but not yet commercially available
Carbon capture	Chilled ammonia capture process Amine scrubbing	Post-combustion capture Oxyfuel combustion Pre-combustion capture
Geologic storage	CO_2 injection with oil or methane recovery Geological monitoring and modeling methods for CO_2 fate and transport	Saline formations Deep seam coal beds
Terrestrial sequestration	Cropland, forestland, and grazing management with advanced information technologies	Genetic engineering to enhance biological carbon uptake

Source: Modified from Brown, M. A., & Sovacool, B. K. (2011) *Climate Change and Global Energy Security: Technology and Policy Options.* Cambridge, MA: MIT Press, p. 67.

Capturing and sequestering carbon

To make the transition to a clean coal economy possible, large volumes of carbon will have to be captured and sequestered. In this regard, proponents of clean coal argue that the technical capacity currently exists, with commercially available or technically feasible options summarized in table 7.2. CO_2 is already routinely captured as a by-product of ammonia and hydrogen production and from limestone calcinations. Moreover, engineers have been experimenting with pre-combustion capture, post-combustion capture, and oxyfuel combustion for decades.[55]

Once captured, scientists around the world have been looking for ways to store, rather than emit, CO_2 produced by fossil-fuel combustion and industrial processes. Two very different options for this task are geologic storage and terrestrial/biological storage. Long-term storage of GHGs in geologic formations is one possible way to avoid emissions. Such geologic formations, located deep underground, could store injected CO_2 much like natural gas and oil have been stored naturally for millennia. CO_2 could also be injected into deep ocean masses, or it could be solidified in mineral carbonates, which may be the ultimate approach to preventing leakage back into the atmosphere. Other promising options include trapping and storing CO_2 in depleted oil and gas reservoirs, saline formations, and un-mineable coal seams.[56] Terrestrial or biological sequestration is the conversion of atmospheric CO_2 to carbon stored in vegetation and soils through photosynthesis and the prevention of CO_2 net emissions from terrestrial ecosystems into the atmosphere.[57] Compared to geologic storage, which is typically envisioned at point sources of CO_2, terrestrial sequestration is less limited to specific locations and is widely available as a carbon mitigation option. Most of its potential lies in above- and below-ground tree biomass in forests.[58]

Advocates highlight that if society does not develop CCS, it will simply be unable to continue industrial life-as-we-know-it in industrialized countries. For instance, the IEA noted that in the Nordic region, a part of Europe with some of the most aggressive climate and energy policies in the world (where some like Denmark are aiming to be fossil-fuel-free by 2050), achieving those targets is impossible without CCS. If those Nordic countries expect to retain any of their heavy industry, they will need to successfully equip 50 percent of cement plants with CCS by 2050 and 30 percent of iron, steel, and chemical plants. As they put it, "CCS represents the most important option among new technologies for reducing industrial CO_2 emissions after 2030."[59] If combined cycle natural gas power plants also come to replace larger nuclear and coal-fired units, those facilities would also need to develop long-term CCS systems.

However, critics of CCS argue that very little research has been done to verify the capacity of clean coal solutions to resolve energy security problems in a safe and economically viable manner.[60] Misplaced confidence in clean coal only exacerbates the problem because as the debate continues, so too do the unsustainable consumption practices that have given rise to climate change. Proponents of this view argue that the rule of thumb should be "keep coal in the ground" and avoid diverting funds to these fanciful schemes. Opponents caution that CCS is nowhere near being commercially developed at scale, and that decarbonization scenarios that rely heavily on breakthroughs in CCS therefore hinge on an element that is heavily uncertain, which is risky and likely will shift the burden to future generations.

Socio-technical barriers to change

Even though we have made great progress in terms of understanding the forces behind climate change and energy insecurity and assessing available solutions to them, a mesh of obstacles and impediments toward effective policies and actions remains. The barriers facing climate-friendly technologies are tenacious, interconnected, and deeply embedded in our social fabric, institutional norms, and modes of production across the world.[61] A thorough understanding of these impediments provides a basis for developing effective strategies to accelerate technology commercialization and deployment.

This chapter classifies barriers synthetically and a bit differently, by merging together social and technical aspects, and by placing barriers into the four categories of cost-effectiveness, policy, intellectual property, and socio-cultural domains.

Cost-effectiveness barriers

The commercialization and deployment of technologies is largely a private-sector activity to gain market advantage, ultimately leading to increased profits. Consumers are not likely to adopt low-carbon technologies and practices that are not cost-competitive in the absence of policies or incentives.

Uncertainties associated with the production costs of new products and the possibility that a superior product might emerge are two of the reasons why firms generally focus on their existing competencies and away from alternatives that could make their present products obsolete. Capital investments in firms go preferentially toward perfecting the performance and reducing the production costs of existing products. This technology "lock in" phenomenon, mentioned above in the section on energy transitions, helps to explain why new enterprises and not incumbent firms are typically the source of radical innovations that displace existing dominant designs. Lock-in is also reinforced by financial institutions, which prefer to make loans to companies with collateral and the ability to repay debts – characteristics of successful firms within the existing network.[62]

Types of cost-effectiveness barriers that impede market introduction and penetration include unpriced externalities, high costs, and technical and market risks. Externalities can make it difficult for clean energy technologies to compete in today's market where GHG emission reductions have only limited market value. For example, the price of a ton of carbon dioxide traded on the European market for credits has dropped below €5 repeatedly. Clean energy technologies also often have inherently higher up-front costs due to the need for additional features and subsystems required to achieve GHG reductions. Additional features or systems can increase the capital to operating expense ratio. For example, SF_6 is a high Global Warming Potential gas used in the magnesium industry as a cover gas. SO_2 is being considered as an alternative, but it is more toxic and therefore requires additional monitoring (and cost) to deal with the health and safety issues.

High costs mean that some combination of the capital cost of the technology, its cost of operations, or other aspects of a project that employs the technology yields a product that costs too much relative to other options that perform essentially the same function. High costs of a technology deter investments, making it difficult to justify providing capital to the high-cost technology or financing the use of its outputs in the absence of deployment assistance. High costs, whether real or merely perceived, are impacted by market and technical risks associated with commercialization or commercial deployment of a technology. The high cost barrier is a function of endogenous costs (e.g., the nature of the fabrication process and its materials requirements), but it also reflects fiscal and regulatory uncertainties: the intermittent nature of tax credits and the lack of approved permitting procedures for offshore wind development are two examples from the United States, and the sudden cancellation or reversal of government support for (separately) CCS, zero-carbon homes, and solar PV in the United Kingdom are three examples from Europe.

Policy and regulatory barriers

Unfortunately, just as markets can fail to adequately price energy services and the benefits of low-carbon systems, so can public policies designed to correct them. Government action has its own set of problems, too. Public policies can provide broad societal benefits that increase overall economic welfare, for example, but can also inadvertently disfavor certain segments of the economy, including, in some cases, inhibiting the commercialization and deployment of clean energy technologies.

Competing priorities also arise as a result of legal inertia. For example, it takes a long time to adopt and modify regulations; as a result, they can be slow to adapt to technology advances and therefore inhibit innovation. China, in particular, only implements major economy-wide plans in five-year increments. Similarly, environmental standards that propelled the large-scale reduction of acid rain in the 1980s now enable the continued operation of some of the most polluting power generators in the US far beyond their normal life and dis-incentivize investing in plant upgrades. Competing policies can be caused by outdated fiscal and tax rules: for example, back-up generators (which provide reliability at the expense of energy efficiency and clean air) are depreciated over three years, while a new combined heat and power system (providing both reliability and energy efficiency) is depreciated over 20 years.[63]

Competing regulations are another barrier. Common examples of regulation in the energy sector include attempts to control market entries, prices, wages, pollution, and standards of production and performance. Venezuela, for example, controls the market price of petroleum and gasoline so that it is artificially low. There, the price for gasoline is a mere 6 cents per gallon (or 1.6 cents per liter) – less than one-fiftieth of what somebody in California pays – perhaps explaining why gasoline consumption there is 40 percent greater than any other Latin American country.[64] Other regulatory barriers that disadvantage clean energy technologies and impede efficient market functioning are called "competing regulations" because they favor conventional energy sources and discourage technological innovation. They include a range of environmental performance standards, power plant regulations, rules impacting the use of combined heat and power, parts of the federal fuel economy standards for cars and trucks, and certain codes and standards regulating the buildings industry. Burdensome and underdeveloped regulations and permitting processes can also inhibit low-carbon technologies.[65]

Intellectual property barriers

Generally, intellectual property law is intended to stimulate innovation, entrepreneurship, and technology commercialization. However, its application can also impede the innovation process. For example, patent filing and other transaction costs associated with strong patent enforcement and protection, as well as the anti-trust challenges related to technological collaboration and patent manipulation, can be serious barriers to technology diffusion. Anti-competitive business practices can also play their part in impeding cleaner energy systems.

Patent warehousing, suppression, and blocking are anti-competitive practices undertaken by incumbent firms that impose barriers to technological change.

- **Patent warehousing** is a form of patent manipulation that involves owning the patent to a novel technology but never intending to develop the technology.[66]
- **Patent suppression** involves refusing to file for a patent so that a novel process or product never reaches the market. For example, in 1977, Tom Ogle developed an automotive system for Ford Motor Company that used a series of hoses to feed a mixture of gas vapors and air directly into the engine. Ford built a small number of prototypes that averaged more than 100 miles per gallon at 55 miles per hour (2.35 liters/100km), but the technology was ultimately suppressed.[67]
- **Patent blocking** occurs when firms use patents to prevent another firm from innovating. For example, while Ford has used Toyota technology (in the Ford Escape), Ford has resisted purchasing Toyota's technology for hybrid vehicles because of hefty licensing fees, and likewise, Honda has not been able to successfully negotiate a license to use nickel metal hydride batteries in their hybrid vehicles. General Electric has also used its patent on variable speed wind turbines to prevent Mitsubishi (a Japanese firm) and Enercon (a German firm) from entering the US market.[68]
- **IP transaction costs**. High patent filing costs can serve as a financial impediment for inventors and firms with scarce capital, including many small businesses. While the transaction costs associated with patent filing will vary depending on the type of technology and breadth of the patent, Quinn estimates that typical costs range anywhere from US$10,000 (€7,740) to hundreds of thousands of dollars per patent.[69]

These expenses do *not* include the costs associated with patent continuation, maintenance, and enforcement against infringement. Potts calculates that based on the uncertainties of the patent search process and the number of amendments and drawings that may be required, inventors and firms can expect to expend an additional $20,000 (€15,478) for each foreign country in which patent protection is sought. Furthermore, the patent filing process typically takes between 24 and 36 months.[70]

Moreover, inconsistent or nonexistent patent protection in developing countries and emerging markets can deter innovation, as firms believe they would be at a competitive disadvantage to distribute their technology. Also, many companies do not want to collaborate with overseas partners because participation may attract those that have the most to gain and the least to contribute, risking an asymmetrical relationship where sharing is uneven between firms. Furthermore, host companies in developing countries may be reluctant to purchase or acquire technology that they believe competitors could freely copy in their own markets.

Weak international patent protection affects both the supply and demand components of technological diffusion. Such barriers are especially true for efficient industrial boilers, fluidized bed combustion, coal gasification, and various end-of-pipe pollution abatement technologies such as CCS. For example, weak intellectual property protection has discouraged US and European companies from developing more advanced clean coal technologies (such as more efficient coal washing processes, advanced combustion turbines, and CCS systems). Intellectual property concerns connected with clean coal systems are cited as one of the most significant impediments toward diffusing such technologies to China, Indonesia, and other developing countries – especially where new technologies could be reverse engineered or copied.[71]

Sociocultural barriers

Many additional barriers inhibit the deployment of GHG mitigation technologies in ways that are not captured by the categories discussed thus far. These barriers stem in part from the cultural traits that impact the behavior and choices of individuals. The influence of lifestyle and tradition on energy use is most easily seen by cross-country comparisons. For example, cold water is traditionally used for clothes washing in China, whereas hot water washing is common in the US and Europe.[72] Similarly, there are international differences in how lighting is used at night; the preferred temperatures of food, drink, and homes; practices of bathing and cooking; and the operating hours of commercial buildings.[73]

Energy (and climate) is also a situation where much misinformation abounds, leading to consumers and citizens who are ambivalent or hostile to new innovations, given they fundamentally do not grasp energy and climate issues. In the US, one survey undertaken by Southern California Edison in the 1970s asked thousands of customers where electricity came from, and most people replied "out of the plug in the wall" while others even said "lightning" and "static electricity."[74] In another poll from the 1990s, only 39 percent of those interviewed had ever heard of GHG emissions and of those who had, more than half did not know where they came from.[75] In 2004, 41 percent of respondents in a Kentucky survey identified coal and oil as "renewable resources"[76] And a national survey found that four-fifths of electricity customers were unable to name a single source of renewable energy.[77] A separate study found that nearly 70 percent of flexible fuel vehicle owners (people who purchased automobiles that could run on gasoline and/or ethanol) were unaware that they were driving one.[78] A survey of 1,300 American residents noted that 80 percent thought that idling their vehicles for more than 30 seconds was better than restarting engines when in reality it wastes fuel and emits more carbon dioxide per year than the entire American iron and steel manufacturing sectors.[79] Another study of drivers found that no respondents analyzed vehicle fuel costs in a systematic way, almost none tracked gasoline costs over time, and few considered transportation fuel costs in their household budgets; the study concluded that drivers "lack the basic building blocks of knowledge" necessary to make intelligent decisions about driving.[80] People in the US are so uninformed about energy that a profitable scam consisting of selling a "solar clothes dryer" for $30 and then mailing them a piece of rope and some clothes pins has run for decades.[81]

Public misunderstanding does not solely exist in the US. In Europe, somewhat novel efforts to make energy visible – such as a "Power Aware Cord" that glowed with greater intensity the more energy that was flowing through it – resulted in increased energy consumption because people found the colors aesthetically pleasing.[82] Almost two-thirds of people in Belgium do not know their heating bill and half do not know their electricity bill.[83] In Denmark, a majority of the sample of highly educated survey respondents failed a basic energy literacy test.[84] In the Chinese province of Liaoning, an assessment of electricity use at homes found that only 2 percent reported understanding ways to save energy and 40 percent indicated they did not remember their electricity bill.[85] A global assessment of patterns of energy use and transportation choices among 17,000 consumers in a total of 17 countries confirmed that overall literacy of energy issues is low.[86]

Finally, misplaced incentives occur when the buyer or owner of a technology is not its consumer or user – a phenomenon that is referred to as the principal/agent problem in the economics literature. In general, this problem occurs when one party (the agent) makes decisions in a given market, and a different party (the principal) bears the consequences of those decisions. Such market failures were found by Prindle in an assessment for the IEA to be significant and widespread in many energy end-use markets in both the US and other IEA countries.[87] In many market situations, buyers purchase equipment on behalf of consumers without taking into account their best interests.

The following are a few examples of misplaced incentives that inhibit energy-efficient investments in low-carbon technologies:

- In many countries, the energy bills of hospitals are paid from central or federal funds while investment expenditures must come from either the hospital itself or from a local government agency.[88]
- Architects, engineers, and builders select equipment, duct systems, windows, and lighting for future building occupants who will be responsible for paying the energy bills.
- Landlords purchase the air conditioning and major appliances and decide on major renovations, while tenants often pay the energy bills.
- Industrial buyers choose technologies that manufacturers use in their factories.
- Specialists write product specifications for military purchases.
- Vehicle fleet managers select the cars and trucks to be used by drivers.
- New car buyers determine the pool of vehicles available to buyers of used cars.[89]

The involvement of intermediaries in the purchase of energy technologies limits the ultimate consumer's role in decision-making and leads to an under-emphasis on life-cycle costs. Nearly one-third (32 percent) of US households rent their homes.[90] Similarly, 51 percent of privately-owned commercial buildings are rented or leased.[91] For these segments of the market, landlords have a powerful influence over the energy efficiency of the building structures and their equipment.[92] The landlord–tenant relationship is a classic example of misplaced incentives. If a landlord owns the energy-using equipment while the tenants pay the energy bills, the landlord is not incentivized to invest in efficient equipment unless the tenants express their self-interest. Thus, the circumstance that favors the efficient use of equipment (when the tenants pay the utility bills) leads to a disincentive for the purchase of energy-efficient equipment.

Energy transitions and transformations: Fast or slow?

As a result of the range of technological dynamics, performance attributes, and consumer and policy uncertainties discussed in this chapter, substantial debate exists within scholarship as to whether future energy transitions can be fast and accelerated, or slow and protracted. One of the most influential protagonists in favor of the view that transitions are inherently slow is the geographer Vaclav Smil. Smil notes that, globally, coal only surpassed the 25 percent mark of total primary energy supply in 1871, more than 500 years after the first commercial coal mines were developed in England. Crude oil surpassed the same mark in 1953, about nine decades after Edwin Drake drilled the first commercial well in Titusville, Pennsylvania, in 1859. Hydroelectricity, natural gas, nuclear power, and "other" sources such as wind turbines and solar panels *still* have yet to surpass the 25 percent threshold. Smil adds that steam engines were designed in the 1770s, but didn't take off until the 1800s, and the gasoline-powered internal combustion engine, first deployed by Benz, Maybach, and Daimler in the middle of the 1880s, reached widespread acceptance in the United States only in the 1920s, even later for Europe and Japan. As Smil deduces from these examples:

> Energy transitions have been, and will continue to be, inherently pro-longed affairs, particularly so in large nations whose high levels of per capita energy use and whose massive and expensive infrastructures make it impossible to greatly accelerate their progress even if we were to resort to some highly effective interventions.[93]

A further argument supporting the contention that desirable energy transitions are so difficult to achieve relates to the momentum the existing system exerts on actors. In the case of energy systems, large sums of labor, capital, and effort are "sunk" into them that they create their own "path dependency" or "inertia."[94] On top of that, institutional legacies protect the status quo, and political regulations, tax codes, and even banks and educational institutions come to support a particular energy pathway, along with associated coalitions.[95] The end result is that energy transitions, breaking out of these embedded systems, require a "long-term transformation" that is "a messy, conflictual, and highly disjointed process."[96] Collectively, these technological and behavioral forces "lock" us into a carbon-dependent energy system that highly resists change.[97] In order to counteract this inertia, scholars looking at energy transitions have argued that truly "transformative change" must be the result of alterations at *every* level of the system, simultaneously. That is, one must alter technologies, political and legal regulations, economies of scale and price signals, and social attitudes and values together, making transition a grueling process.[98]

That said, a counterpoint to such pessimistic thinking is that while past transitions may have taken a great deal of time, we can learn from them so that *future* transitions are much quicker. More modern energy transitions can be influenced by endogenous factors within a country, like aggressive planning from stakeholders in Denmark leading to a doubling of combined heat and power in less than five years, or the adoption of 40 million new clean cookstoves in Indonesia in 3 years, or the phasing out of coal in Ontario, Canada, in about a decade.[99] As Harvard energy scholar Kathleen M. Araujo has argued:

> Countries can, in fact, alter their energy balance in a significant way – stressing low carbon energy sources – in much less time than many decision-makers might imagine. Critical substitution shifts within [Brazil, France, Denmark, and Iceland] were accomplished often in less than 15 years. Moreover, these transitions were effectuated even amidst circumstances at times involving highly complex energy technologies.[100]

According to recent scenarios, decarbonization definitely seems to imply rapid, simultaneous transformations across several low-carbon systems and technologies. The International Renewable Energy Agency (IRENA) reports in its most recent outlook that between 2015 and 2050, the share of low-carbon electricity in total final energy consumption needs to double.[101] The number of electric vehicles (EVs) needs to jump from almost one million to one billion cars (more precisely from 1.24 million passenger cars to 965 million passenger cars); from 200,000 electric buses and trucks to 57 million; and from 200 million electric scooters and bikes to 2.16 billion.[102] The amount of battery storage similarly needs to climb from 0.5 gigawatt hours (GWh) to 12,380 GWh.[103] The number of heat pumps in households needs to jump from 20 million to 253 million.[104] The amount of installed solar PV capacity must rise from 223 gigawatts (GW) to 7,122 GW.[105] A report from the consulting company McKinsey assessed these trends, and deduced that "energy companies should be planning for an industrial revolution driven by renewables" and that "by 2035, renewables (solar and wind) will account for more than 50 percent of global power generation; electric vehicles will be the low-cost option for car, van and small-truck drivers; oil demand will be declining; and gas demand will have peaked."[106]

This suggests that while history may be instructive about the speed and pace of future transitions, it may not necessarily be predictive. Indeed, when one currently considers whether energy transitions can be fast or slow, the current transformation underway across various sectors, including electricity and transport, does give rise to grounds for hope.

Conclusion

To conclude, a host of technologies relating to energy end-use infra-structure on the demand side, energy supply, and carbon capture and storage are currently available and technically feasible. However, the unique barriers faced by different types of technologies highlight the fact that specific deployment policies and programs may be required. Broadly speaking, barriers hinder low-carbon technology commercialization and deployment in different ways: by locking in incumbent technologies, by escalating the business risks of innovation, or by increasing transaction costs associated with change. These powerful and restraining influences reinforce one another.

At the same time, economy-wide actions may be more efficient in addressing common barriers in a broad, systematic fashion in ways that could significantly accelerate and expand the uptake of clean energy technologies. This tension between highly specific versus general policy interventions requires careful consideration. It also reveals why energy transitions remain such contested and highly political processes.

8

National and regional energy policy

This chapter discusses national energy governance through the lens of policies, a catchall term that refers to plans and regulations as well as strategies, guidelines, goals, and laws, at many scales of action (local, national, global, etc.). After a short primer on what energy policy is, and how it is generally conducted, the national energy systems and policy orientations are discussed of key players, namely the United States, the European Union, China, India, Russia, Brazil, and South Africa. These seven cases were selected to have a balance between established and emerging powers, major demand centers and major exporters of energy, and geographic location. Our goal here is to present the national particularities of the energy situation in a concise and condensed way, showing how each country – or region in the case of the EU – grapples with the competing frames and trade-offs involved in energy policymaking.

Characterizing national energy governance

Governments at national and local levels play a crucial role in shaping energy systems. As the IEA has pointed out repeatedly, "governments hold the key to changing the mix of energy investment."[1] In this section, we want to highlight the goals of energy policy, the challenges to crafting energy policy, the basic instruments to implement energy policy, and the international and domestic factors that shape the decision-making process on energy.

The goals and instruments of energy policy

Energy policy is often ill-defined. What we call "energy policy" today usually refers to a plan of action focused on ensuring sufficient supplies of affordable fuels and electricity to satisfy market demand and, increasingly, to do so while shifting to low-carbon energy sources in the face of climate change.

However, this narrow definition fails to recognize the permeable boundaries of energy policy, shaping and being shaped by fiscal policy, foreign policy, social and health policy, science and technology policy, climate policy, and other government concerns. The energy sector is shaped to no small extent when governments act to pursue non-energy objectives – for example, curbing inflation, stimulating employment, spatial planning, regional development, national security, poverty alleviation, and carbon emissions reduction.[2] The opposite is also true: energy policy decisions can affect all sectors of the economy, including transportation, business, and trade.

There are many challenges to crafting sound energy policies. One of the problems for governments is that politicians in democratic societies have short time horizons, often looking no further than the next election, whereas many energy and climate problems play out over the long term but require action today.[3] This problem is accentuated today by the rise of right-wing populist parties in the industrialized world and elsewhere. One of the key tenets of contemporary (right-wing) populism is a profound distrust of experts and elites. Multilateral climate accords such as the Paris Agreement are often portrayed as part of the "establishment" that populist voices are keen to overthrow.

Another problem often encountered by governments is institutional fragmentation. As said, energy cuts across different policy domains and cannot be housed in a neat box. Energy is therefore often placed in a department together with other policy domains, most often economic affairs or the environment (see table 8.1).[4] Some countries subdivide responsibility for energy into its different sources. For example, control over India's energy policy and planning is spread over five fuel-based ministries: coal, power, petroleum and natural gas, new and renewable energy, and atomic energy.[5] Other countries, by contrast, have no departments of energy at all – at least, no national ministry or department that mentions energy in its name. This is the case in China and Japan, for instance.[6] Even if a department or ministry exists, it may not have much authority, like in the case of the US Department of Energy, which

Table 8.1 Energy in ministerial portfolios across selected countries

Country	Main government department(s) responsible for energy
Australia	Department of the Environment and Energy
Austria	Federal Ministry for Sustainability and Tourism
Brazil	Ministry of Mines and Energy
China	Ministry of Natural Resources and Ministry of Ecology and Environment
France	Ministry of Ecology, Energy, Sustainable Development and Spatial Planning
Germany	Federal Ministry for Economic Affairs and Energy
India	Ministry of Power, Ministry of Petroleum and Natural Gas, Ministry of Coal, Ministry of New and Renewable Energy, and Department of Atomic Energy
Japan	Ministry of Economy, Trade and Industry
Norway	Ministry of Petroleum and Energy
Spain	Ministry for the Ecological Transition
United Kingdom	Department for Business, Energy & Industrial Strategy (formerly the Department of Energy and Climate Change, or DECC)
United States	Department of Energy

Source: Government websites of respective countries, accessed May 2019.

is generally regarded as one of the weaker US federal government departments.[7]

For many countries in the world, the goals of energy policy reflect a trilemma, involving the (sometimes) competing demands of energy security, climate change mitigation, and economic development.[8] These energy dilemmas play out differently in different regions and countries. In the high-energy societies of the developed world, there is a strong and growing tension between the climate-change imperative of decarbonization and the affordability dimension of energy security. For the emerging economies, the imperative to secure sufficient energy to continue to fuel economic growth often takes precedence over concerns about emissions. In the developing world, finally, issues of energy access take clear priority over the promotion of clean energy.[9]

When implementing energy policy, governments can employ a number of policy tools. Generally, two sets of policy tools can be identified. *Command-and-control approaches* impose direct regulation on the energy industry. An example is the use of performance standards, which includes regulations on how much CO_2 cars may emit or how energy-efficient household appliances or buildings should be. It may also involve prohibition of certain products and practices, such as a ban on the sale of leaded gasoline or a phase out of nuclear power. Renewable energy quota obligations also fall into this category. *Market-based approaches*, by contrast, influence actors' behavior by changing their economic incentive structure. Putting a price on carbon – whether through a carbon tax or a cap-and-trade system – is an example of such an economic incentive. Product labeling and public information campaigns are also compatible with free markets, but represent an even softer form of governance. Whereas economists have displayed a clear preference for market-based approaches like carbon pricing to bring about a clean energy transition, political scientists have suggested that direct regulatory and fiscal support to green industries can help build winning coalitions for decarbonizing national economies.[10]

Drivers of national energy policy

Energy policy trajectories are undoubtedly influenced by natural conditions, such as topography (e.g., availability of hydropower potential), occurrence of natural resources (e.g., availability of oil), climate (e.g., different heating and cooling loads of buildings), geography (e.g., proximity to a country with gas export potential), and population density (e.g., space available to deploy energy assets such as windmills, fracking installations, or nuclear power plants).[11] The state of technology and its cost will undoubtedly also inform energy policy – it is likely, for instance, that the strong drop in costs of several low-carbon technologies created new economic realities that, in turn, helped to foster success at the Paris climate summit in 2015.[12]

These material factors – the resource, technical, and financial endowments of a particular country – are important drivers of national energy policies. They establish, in the words of John Duffield, "the outer bounds of possible policy responses" to particular energy problems and challenges.[13] However, we cannot fully understand the evolution of national energy policies without an appreciation of the critical role played by political factors.[14] The political drivers of national energy policy can be summarized as ideas, interests, and institutions.[15]

With regard to *ideas*, a key factor is the dominant view of what should be the role of the state in the economy. The energy sector has not escaped the marked swings in the state–market pendulum that has characterized the broader political economy.[16] In the first decades after World War II, a statist approach prevailed, whereby energy sectors such as electricity and gas were often administered by state companies or regulated monopolies. In the 1980s, the free market paradigm took over, making energy subject to private provision instead. This was the period when the power gas sectors were liberalized in many parts of the world, just like telecommunications and railways. More recently, the global political economy has returned to an era of state capitalism or interventionism. One illustration of this trend is the growing appeal of so-called "green industrial policies," in which the state assumes a direct role in promoting certain low-carbon technologies and sectors.[17]

Regarding *interests*, a distinction can be made between public and private interests. An example of a public interest that informs the energy policies of most countries is energy security. The key problem, however, is that there is no commonly accepted definition of energy security, which opens the door for actors to justify just about any possible action or policy on "energy security" grounds.[18] The same goes for other public goals such as "sustainability" or "energy independence."[19] Various interest groups will seek to influence energy policy, in pursuit of public, group, or private interests. These interest groups range from social movements (e.g., the "Youth for Climate" or "Fridays for Future" school strikes or the "*gilets jaunes*" protesters), over organized non-governmental organizations (e.g., Greenpeace, 350.org), to industry associations (e.g., farmers, fossil fuel companies, renewable energy companies). Such groups sometimes try to directly influence policy processes, through lobbying, but they can also weigh on policy indirectly, by building coalitions or shaping public discourses.[20]

The space that these various actors and interest groups have to influence energy policy varies hugely by country. It depends on the *institutional* structure of the state. One of the key differences concerns state capacity and autonomy. Some countries have a strong bureaucracy, while weaker states often have less policy autonomy and institutional capacity to steer society in a particular direction.[21] Another line of research, the varieties of capitalism literature, has emphasized cross-national differences in state–market relations. In liberal market economies, like the UK, the US, and Australia, the state plays a minimalist role in the economy. In coordinated market economies, like Denmark, Sweden, and Germany, the state typically assumes a more coordinating role.[22] Research has suggested that the latter type have a better ability to address climate change.[23] A third distinction is between political systems with centralized or diffused political authority. Existing studies have focused on the differences in energy policymaking processes and outcomes between democratic and autocratic regimes, presidential and parliamentary systems, and federalist and non-federalist states.[24]

A final key driver of energy policy concerns temporality, or the impact of earlier policy decisions on later ones. Energy policy often develops reactively, in response to a particular shock or crisis. As cases in point, the energy accidents at Chernobyl and Fukushima both culminated in an aggressive diversification of electricity portfolios away from nuclear power; the oil shocks of the 1970s had a direct link to the Danish push for wind power and coal (to diversify away from oil) as well as Brazil's national ethanol program (to reduce oil imports).

But policy decisions taken to address a short-term issue may have longer-term consequences. It may set a country on a particular path that reinforces itself through mechanisms such as lock-in and positive feedback effects. As a result, once a particular policy option is selected, it might progressively become more difficult to go back to the initial point when multiple policy alternatives were still available. This effect is known as "path dependency."[25] The effect of such policy legacies is amplified by the energy sector's sheer scale and capital intensity, as mentioned in chapter 4. Energy policies often promote investments in particular types of energy production, consumption, or transportation that require years and sometimes even decades to amortize.

Energy policy in selected countries

Of course, the above discussion is largely theoretical or at least conceptual. The remainder of this chapter shows the actual empirical use and strategy of energy policies in the real world, with an examination of the three largest energy consumers and emitters of carbon dioxide (the United States, European Union, and China) as well as the BRICS countries (India, Russia, Brazil, and South Africa). Collectively, this group of countries comprises more than half of the world's population and they account for about two-thirds of global GDP, global energy demand, and global CO_2 emissions (see table 8.2). Better comprehending their energy policy dynamics, and their strategic energy concerns, may hold the key to a future sustainable global energy transition.

To avoid each case from becoming excessively long, we have limited our focus to recent history, that is, energy policymaking and planning from the 1970s to today.

Table 8.2 Descriptive economic, demographic, and energy statistics for the selected country policy case studies

	Share in global GDP (2018)	Share in global population (2018)	Share in global energy demand (2016)	Share in global CO_2 emissions (2017)
Brazil	2.2%	2.8%	2.1%	1.3%
China	15.9%	18.3%	21.7%	27.2%
European Union	21.9%	6.8%	11.2%	9.8%
India	3.2%	17.8%	6.2%	6.8%
Russia	1.9%	1.9%	5.3%	4.7%
South Africa	0.4%	0.8%	1.0%	1.3%
United States	23.9%	4.3%	15.9%	14.6%
TOTAL	69.3%	52.6%	63.5%	65.7%

Note: EU includes UK.

Sources: World Bank, World Development Indicators; IEA (2018) *World Energy Balances*; Boden, T. A., Marland, G., & Andres, R. J. (2017) Global, Regional, and National Fossil-Fuel CO_2 Emissions. Carbon Dioxide Information Analysis Center, Oak Ridge National Laboratory, US Department of Energy, Oak Ridge, TN; UNFCCC (2018) National Inventory Submissions 2018; BP (2018) *Statistical Review of World Energy*. Available at http://www.bp.com/en/global/corporate/energy-economics.html.

United States

Energy became an issue of real public concern for the United States only in the 1970s. The 1973 oil embargo, which targeted the US specifically, was an important wake-up call about the US's growing dependence on imported oil. Several weeks after the beginning of the embargo, President Nixon gave a televised address, in which he proclaimed a long-term objective: "Let us pledge that by 1980, under Project Independence, we shall be able to meet America's energy needs from America's own energy resources."[26] The goal was never reached, yet "energy independence" would become a goal of every US president since Nixon.[27]

Nixon's most important energy policy measure had already been taken before the Arab oil embargo, namely to put in place oil price controls to curb rampant inflation. The 1975 Energy Policy and Conservation Act (EPCA), adopted under President Ford, introduced mileage standards for automobiles and established the Strategic Petroleum Reserve.[28] When Carter took office, he made energy one of his key priorities, and put out a plan with more than 100 concrete proposals. Congress was only able to pass five energy bills, put together as

National Energy Act of 1975. The Act mainly addressed conventional fossil fuels as it attempted to move the country toward energy independence, promote the use of coal, and stimulate conservation.[29] Carter's final two years in office were perturbed by the 1979 oil shock and nuclear accident of Three Mile Island.

Among the set of five laws proposed by Carter and passed by Congress (albeit in greatly diluted form), the Public Utility Regulatory Policies Act (PURPA) of 1978 had far-reaching consequences for small-scale energy systems. PURPA spurred creation of radical technologies, began the process of deregulation, and challenged the control held by power company managers. PURPA offered incentives for the use of decentralized power plants (such as wind farms and solar panels), and spurred research on environmentally preferable technologies that used water, wind, or solar power to produce electricity. More successful than anyone originally anticipated, PURPA established the first production tax credits for renewable energy systems such as wind turbines and solar panels.[30]

When Reagan took office in 1981, he took a decidedly different approach to energy policy, and indeed, the economy in general. He wanted to roll back the role of government in the economy and deregulate energy markets. In one of his first acts in office, Reagan abolished the oil price controls that Nixon had put in place. His administration also slashed the budgets for energy research and development (R&D) and other renewable energy programs that had been set up in the 1970s.[31] Reagan also attempted to abolish the Department of Energy (DOE), which had been created in the 1970s, but was not able to do so due to opposition in Congress.[32]

The next major Act came in 1992, the Comprehensive National Energy Policy Act. While much of this bill focused on energy efficiency and utilities, it had in fact something for every fuel. It had some provisions that made the expansion of nuclear power easier, and it introduced tax credits for electricity from wind, solar, biomass, and geothermal facilities. These tax credits have sometimes lapsed since 1992, but have nevertheless become a permanent feature of US energy policy.[33]

It took another 13 years for the next major revision of energy laws, which came in the form of the Energy Policy Act of 2005. Like its 1992 predecessor, this Act sought to promote every form of domestic energy, from fossil fuels to nuclear power to renewables. The hodgepodge of technologies supported by the act even provoked *The Economist* to call it "the no lobbyist left behind bill." Yet, the biggest winners were arguably the biofuels sector, as the Act mandated a doubling in the nation's supply of ethanol or biodiesel by 2012, as well as the nuclear sector, with $2 billion in federal support for construction overruns; a production tax credit worth 1.8 cents/kWh; loan guarantees of up to 80 percent of project cost; $3 billion in new research; updated favorable tax treatment of decommissioning; and the extension of limited liability for nuclear accidents under the Price Anderson Act.[34]

The Energy Security and Independence Act, passed two years later in 2007, was not as wide ranging and sweeping, but it did include stricter standards for fuel economy for automobiles, additional support for biofuels, and more commitment to energy efficiency and buildings and demand-side management. Notable, here, was renewed support for biofuels with a requirement that production grow from 4.7 billion gallons in 2007 to 36 billion gallons by 2022 and that 21 billion of these gallons be derived from non-corn-based materials. The Act also revised federal labeling and appliance standards for compact fluorescent light bulbs (effectively aimed at dis-incentivizing the use of incandescent light bulbs) and mandated that all federal buildings become carbon neutral by 2030.

President Obama continued the expansive view of energy policy, repeatedly stating that his energy policy was an "all-of-the-above" strategy, emphasizing all forms of energy. At the same time, he attempted to promote clean energy, which got a big push in the economic stimulus plan adopted in his first term. President Trump, conversely, has focused his first term in office (so far) on withdrawing the United States from the Paris Accords; on increasing subsidies and federal incentives for oil, gas, and coal; and on relaxing pollution standards and other targets such as the Clean Power Plan.

European Union

The European Union (EU) has developed one of the world's most ambitious climate and energy policies. Its 28 member states have jointly committed to the famous "20-20-20" goals, which involve a reduction in greenhouse gas emissions by 20 percent, an increase in energy efficiency by 20 percent, and an increase in the share of renewables in the final energy mix to 20 percent, all to be realized by 2020. These goals have been further strengthened for 2030. Furthermore, the EU has long assumed a leadership role in international climate diplomacy, and several European companies have been early movers in the development of green industries. In addition, the EU operates the very first and most advanced emissions trading system in the world.

And still, energy represents somewhat of a paradox of EU integration. The whole European integration process once started with energy, the European Coal and Steel Community (1952) and Euratom (1957), but the engine of integration in the energy field sputtered after the adoption of these treaties. After the first oil shock of 1973, the Americans took the lead in setting up a new international organization, the International Energy Agency (IEA), which was housed at the OECD in Paris. Henceforth, the energy policies of the European member countries would be discussed and coordinated in Paris, at the IEA, not in Brussels.

Things would begin to change with the "1992 project" of completing the Single European Market. It would kick start the liberalization of gas and electricity markets, which long remained the hallmark of the EU's energy policy. The EU's energy policy has long derived from the internal market competences, but also competences with regard to transport and environmental policy. While the Treaty of Lisbon (2007) has finally provided a legal basis for the EU to develop a proper energy policy, the energy mix remains the exclusive prerogative of the member states. In practice, however, this provision amounts to legal fiction. To be sure, national governments in the EU have vigilantly guarded their sovereignty over energy policy. Yet, successive EU energy and climate regulations (e.g., the 20-20-20 goals) have ensured that the space for each member state to autonomously decide on its energy mix has become smaller over time.

The underlying goal of the push to liberalize gas and electricity markets was not related to energy security or to climate change, but to make sure that energy providers operate in a competitive environment that ensures affordable prices for end users (industry, businesses, families). Natural gas and electricity are two network industries characterized by a fixed infrastructure and thus prone to what economists call a "natural monopoly," much like the railroads and telecommunications sectors, other markets which the EU sought to liberalize. Breaking down the old state-led monopolies in these network industries proved a challenge for the EU. The European Commission's key goal was to make transmission operators independent from supplier interests, with the expectation that this would lead to sufficient cross-border infrastructure and generation capacity. Ownership unbundling was the major focus of the legislative packages of 1996–1998, 2003, and 2009.

The so-called Third Energy Package of 2009 was the most comprehensive. The package not only comprised the unbundling provisions, but also the creation of more powerful networks of European regulators – the Agency for the Cooperation of Energy Regulators, or ACER – and of transmission operators – the European Network for Transmission System Operators for electricity and gas, the so-called ENTSO-E and ENTSO-G.

Even though this legislation was quite effective and forceful, the Commission soon realized that it needed to intervene more because the imperatives of climate change needed the EU to add infrastructure faster than markets could provide. Moreover, the Russia–Ukrainian gas crisis of January 2009, much more severe than the short blip in gas supplies in 2006, served as a wake-up call for the EU and showed that the Commission had much more work to do in terms of security of supply measures (ensuring bi-directional flow of gas pipelines, regulating energy security policies, coordinating strategic stocks of gas, stress testing the system, etc.).

When the Juncker Commission took office in 2015, it made the "Energy Union" one of its priorities. Among industrialized states, the EU by then already had one of the world's most ambitious climate and energy policies (the famous 20-20-20 goals). Building on these prior commitments, the EU member states agreed on a new 2030 Framework for climate and energy in October 2014, targeting a 40 percent cut in greenhouse gas emissions compared to 1990 levels, as well as renewable energy and energy efficiency targets of at least 27 percent.

With regard to the emission reduction goals (20 percent by 2020; 40 percent by 2030), it is important to make a distinction between those sectors of the economy that fall under the EU's Emissions Trading System (ETS) and those that do not. The ETS system is the CO_2 emission cap-and-trade system that the EU set up in 2005. Currently, only certain sectors fall under this system: the power sector, industry, and (since early 2012) the aviation sector. Those sectors have to reduce their emissions by 21 percent in 2020 compared to 2005 levels and by 43 percent in 2030 (see box 8.1). To that end, the EU puts a ceiling (or a "cap") on the total allowed CO_2 emissions. Companies can then either try to lower their emissions or they can go to the market to buy CO_2 emission permits (called "allowances" in EU parlance). To date, they can also buy limited amounts of international credits from emission-saving projects around the world, and particularly in developing countries, but the EU does not envisage continuing the use of international credits

---

Box 8.1   The European Emissions Trading System

The flagship project in the 2020 and 2030 strategies is clearly the EU's Emissions Trading System (or "ETS"), the world's first and still by far the largest cap-and-trade mechanism for CO_2, operating in 31 countries (all 28 EU countries plus Iceland, Liechtenstein and Norway). Currently, the system covers in excess of some 11,000 installations, which collectively represent almost half of Europe's CO_2 emitters. This system sets a CO_2 emissions ceiling and puts a price on CO_2 emissions. The basic idea of the cap-and-trade system is that the ceiling of allowed emissions will be lowered each year, so that the price of carbon goes up, and investments in low-carbon energy are stimulated. This is clearly linked with another goal, that of promoting renewable energy. Moreover, there is also a close connection between the renewables and efficiency targets because the renewable energy target is expressed as a share of final energy demand. Lowering demand through increased efficiency thus helps to achieve the renewables target.

The system went through a pilot and second phase and entered its third phase from January 2013. This phase will finish in 2020. With the crucial last phase, a number of changes occurred in the whole system. The most important changes relate to the way in which the permits are allocated: first, there is now an EU-wide cap rather than national caps; second, permits are sold on the market rather than given for free; and, third, the plan is that, each year, the number of permits will decrease with the intended effect of ramping up the low price of carbon, which has been the Achilles heel of the whole process so far (excess supply tackled through back-loading of auctions in phase 3 and "market stability reserve").

As an indicator of just how vulnerable the ETS system had become, in late January 2016, the carbon price fell to under €6 per metric ton of carbon. This is still obviously far below the €20 or €30 price point that analysts say is needed to spur the type of clean investment needed by industry to cut carbon emissions, although as of mid-2019 credits were trading at above €25 again.

---

after 2020. Those sectors that do not fall under the ETS system, such as (land) transportation, buildings, agriculture, and waste, only have to reduce their emissions by 10 percent in 2020 compared to 2005, while in the new framework the aim is emission reduction of 30 percent by 2030.

China

China started on its path toward rapid industrialization in 1978, when a period of economic reform "opened up" China after the chaotic period of the Cultural Revolution. The 1970s and 1980s were characterized by low levels of energy demand, an underdeveloped electricity sector, and a generally lagging economy.[35] Energy shortages were seen as the principal culprit behind these problems, and so Chinese planners aggressive promoted the development of new energy resources, especially those that facilitated industrial growth. The strategy created successful momentum toward both energy and economic growth, and between 1990 and 2010 China's economy grew almost fivefold, and its energy use more than doubled.[36]

It was during the 1990s that Chinese planners yielded a broad package of market reforms to transition from a centrally-planned economic system to a more market-oriented one, a strategy that had relatively profound impacts on the country's domestic energy sectors.[37] Growing dependence on imported energy became a particular concern to Chinese planners in 1993, when China ended three decades of energy self-sufficiency and became a net importer of oil.[38] Rapid Chinese economic growth created a parallel growth in oil and gas demand, necessitating imports from international oil markets and raising concerns about energy efficiency.[39]

One part of the Chinese plan to address security of supply concerns focuses not inward, on further development of domestic resources, but outward, by relying on energy fuels from its acquired foreign assets instead of imports and short-term global energy markets. This was evident in an aggressive "going out" strategy involving overseas oil and gas asset purchases by its state oil companies such as CNOOC (China National Offshore Oil Company), CNPC (China National Petroleum Corporation) and Sinopec (China National Petrochemical Corporation).[40] A focus on overseas markets was coupled with increased attention to domestic markets. The two Five-Year Plans of the 1990s (the Eighth and the Ninth) both heavily supported expansion of electricity supply and energy for industrialization, targeting annual growth rates often as high as 7 to 10 percent, with particular subsidies given to the coal sector as well as long-distance high-voltage electricity transmission networks.[41] This part of the strategy reflected the fact that increases in consumption have outpaced the ability for Chinese planners to secure foreign assets.[42] Several subsequent policies sought to diversify national supply to renewable sources of energy, with major policy initiatives including China's national Agenda 21 in 1992, the Brightness Program and Ride the Wind Program of the State Planning Commission in 1994, and the Ninth Five-Year Plan of Industrialization of New and Renewable Energy formulated by the State Economic and Trade Commission (SETC) in 1996.

In 2002, the Hu Administration attempted to accelerate economic reforms by encouraging market liberalization and attempts at reigning in corruption and environmental degradation.[43] The Tenth Five-Year Plan (2001–2005) articulated the need for price liberalization of energy products, especially electricity, but supply did not grow fast enough. From 1994 to 2011, the mean growth rate in energy production was 5.9 percent but consumption rose 6.2 percent.[44] China's energy sector also faced forces beyond its control during this period, with volatile global crude oil prices and restrictions on intellectual property related to advanced energy equipment in place.[45]

Also, though security of crude oil supply continued to increase over the decade,[46] and diversification towards renewables continued with the Tenth Five-Year Plan for New and Renewable Energy Commercialization Development adopted by the SETC in 2001 and the Renewable Energy Promotion Law in 2003, the total share of renewable energy actually decreased, as its pace of development could not keep up with China's growing appetite for energy.[47] The share of renewable electricity, for instance, fell from 23.4 percent in 1985 to 16.8 percent in 2005. This decline is largely explained by truly massive increases in investments in coal.

China's energy future will most certainly differ from its past, as it grapples with at least three interlinked challenges. First, its growing import dependence. China already imports more than 10 million barrels of oil per day. China's dependence on foreign oil is expected to rise to 80 percent by 2030 and on foreign gas to 42 percent by 2030[48] – a major energy security concern. In addition, China needs to continue its efforts to diversify away from coal, oil, and natural gas to renewable sources of energy such as hydrogen and nuclear, and to become more energy-efficient, another major challenge considering future plans for the dominance of coal.[49] Third, China has exhibited a notable lack of coordination in its energy policies. As Green and Kryman write, "there is no such thing as a unified national Chinese energy or climate policy ... The result is a labyrinthine and complicated governance structure, an energy related bureaucracy becoming increasingly contradictory."[50]

India

India is the fourth-largest energy consumer in the world and one of the top ten emitters of greenhouse gases. Yet, annually, the average Indian citizen uses 11 times less energy than a US citizen, produces 10 times fewer greenhouse gas emissions, and uses 16 times less electricity.[51] With more than 1.3 billion people today, India is projected to have the world's largest population by 2024,[52] and is poised to overtake China as the world's largest energy growth market by the end of the 2020s.[53] Much like in the case of China, the energy choices that India makes will set the pace of the global energy transition.

India is largely dependent on fossil fuel imports to meet its energy demands. Although in 1974 India discovered vast petroleum reserves off its West Coast and domestic production accounted for 41 percent of consumption by 1989–90,[54] crude output started to decline within a few years. India has had to continue importing to meet growing demand.[55]

The 1990s saw relatively major changes both to the state's electricity and oil markets. In 1995, the "Mega Power Policy" was introduced to accelerate investment in massive (1,000–4,000 MW) power facilities, and created a national Power Trading Company. In 1998, the Electricity Regulatory Commission Act created the Central Electricity Regulatory Commission and encouraged the creation of State Electricity Regulatory Commissions to set and regulate prices. This was followed by India's New Exploration Licensing Policy in 1999, which helped attract new investment into exploration, and resulted in the discovery of several major oil fields.[56] Over this time, India came to rely heavily on coal for electricity generation. During the 1990s, India became both the third largest consumer and producer of coal in the world. However, India's electricity sector relied on low-quality coal, rendering coal-fired electricity generation inefficient, meaning more of it was needed per unit of energy than in industrialized countries such as Germany or the United States.[57]

The 2000s saw the Energy Conservation Act passed in 2001, which provided the legal and institutional framework to promote energy efficiency throughout all sectors of the economy, and created the Bureau of Energy Efficiency.[58] That same decade, the 2003 Electricity Act created regulatory publications of electricity tariffs, set permitting standards for renewable energy technologies, and promoted cogeneration and energy efficiency. It also unbundled generation, transmission, and distribution entities, granted non-discriminatory access to the transmission grid, and set purchase obligations for renewable electricity, among other items. This was followed by a National Electricity Policy, passed in 2005, which set the target of universal access to electricity by 2010 (which it did not accomplish), and attempted to eliminate electricity shortages by 2012 (which it did not accomplish). It also tried to increase per capita consumption of electricity to more than 1,000 kWh (which it did not accomplish), stipulated the protection of consumer interests and the financial turnaround of electric utilities.[59] A National Tariff Policy, passed in 2006, tried to strengthen the financial viability of the electricity sector to attract investors, guaranteeing double-digit returns on investment, and a 2006 Rural Electrification Policy attempted to (again) achieve universal access to electricity by 2009/2012 (which it did not accomplish).[60] An Integrated Energy Policy was adopted in 2008 to enable India to meet its development goals and grow the economy. It included components focused on expanding renewable energy, continuing to invest in coal, promoting energy efficiency, and minimizing transmission losses, among others.[61]

While the frenetic pace of energy policymaking in the 2000s did not achieve all or perhaps even most of its intended goals, the energy access policies did have positive effects on lowering the country's dependence on solid fuel and in promoting rural electrification.[62] The share of residential energy use met from the centralized grid jumped significantly over this decade, with increases in income, urbanization, and lower energy prices all driving a shift away from solid and liquid fuels.[63] The share of solid fuels meeting household cooking and heating needs dropped from 42 percent in 1990 to 25 percent in 2010.[64]

Looking to the future, major challenges include continuing dependence on coal-fired electricity, declining reliability, even greater import dependence, and the issue of electricity theft. Most significantly, India's growing economy and population are dramatically increasing demand for electricity, most of which will be generated by coal. With an installed capacity of 189,000 MW in 2010, India already consumes the fifth-largest amount of electricity in the world.[65] Coal-fired thermoelectric power plants produced about 71 percent of the country's electricity in 2011–12, with nuclear, hydropower, diesel, and natural gas making up the remainder.[66] From 2010 to 2040, some projections indicated that India could add a staggering 613,000 MW of coal capacity to the grid, with the state of Gujarat by itself expected to add 40,000 MW.[67] Reliability and uninterrupted supply remains a concern, with Pode suggesting in 2010 that "lack of energy resources could jeopardize not only India's economic progress but also security and strategic interest."[68] As a sign that this warning was apt, in July 2012, India suffered an "unprecedented grid failure" that affected 670 million people – more than half the country's population, or roughly 10 percent of the world's population, in a single blackout.[69] India's import dependence will only grow more stark as well, with the IEA projecting that by 2030, 91 percent of India's oil will likely be imported.[70] A final challenge relates to high "aggregated transmission and commercial losses," a term that describes "lost" power due either to inefficiency or theft from illegal tapping of transmission lines. Nationwide, although recent data is hard to come by, in 2011 these losses were greater than 31 percent, and they created a significant burden on electric utilities, as functionally one-third of their electricity produces no revenue.[71]

Russia

Much energy analysis continues to focus on Russia's external role as an energy superpower, namely its use of energy as a "weapon" against countries such as Ukraine or Russian foreign energy policy toward Europe or China – topics we have covered in chapter 3. Nonetheless, its internal dynamics are just as fascinating and complex. Even throughout the recession and economic stagnation confronting the former Soviet Union during the 1970s, authorities maintained their investments in energy, even though such investments created a "curse" in the 1980s as global oil and gas prices plummeted – hitting Russia hard as energy exports accounted for 80 percent of Soviet hard currency earnings.[72] The 1986 nuclear accident at Chernobyl complicated state efforts to expand the nuclear power sector, the government finding itself "under the unexpected and unaccustomed onslaught of an angry public," forcing it to suspend, cancel, or close more than 50 nuclear reactors.[73] Planners in the late 1980s were concerned primarily with managing the beginning of the fall of the Berlin Wall and simply maintain adequate electricity supply against the backdrop of great social and political upheaval.[74]

Russia performed so well in part due to an abundance of energy resources. Russia, for instance, is resource and energy-product rich, holding the world's largest natural gas reserves (second in production, behind the United States), second largest coal reserves (sixth in production), and seventh largest oil reserves (third in production, behind Saudi Arabia and the United States).[75] Oil and gas contribute to roughly half of Russia's GDP and two-thirds of their export earnings.[76]

The collapse of the Soviet Union ushered in broad political and economic reforms which decentralized state planning. Mikhail Gorbachev came to power on a platform of economic liberalization and instituted market reforms. The Politburo was abolished and calls within and outside the government were made for *glasnost* (transparency and openness) and *perestroika* (rebuilding). A new, more inclusive legislative body – the Congress of People's Deputies – was established to replace the State Commission.[77]

Nonetheless, the collapse of the Soviet Union also brought daunting social and economic problems including hyperinflation, food shortages, and electricity blackouts within Russia and its former republics.[78] The relaxation of central state planning, and monitoring, ostensibly intended to improve efficiency, also led to large-scale embezzlement of state funds and misappropriation of state assets, especially within the energy sector.[79] A weak regulatory framework emerged which allowed well-connected state elites to use "insider trading" and "asset-stripping" to transfer wealth from state energy enterprises into private hands.[80] Economic reforms ended in 1998 with a domestic financial crash, an economic decline which reverberated throughout the energy sector by depreciating the value of capital stock and the strength of the ruble. In the oil and gas industry, average volumes of exploratory drilling thus declined (from 5.3 million meters in 1990 to 1.4 million meters in 1998) alongside drops in production from 569 million metric tons in 1988 to 305 million metric tons in 1999.[81] This convinced the government to begin more aggressively partnering with foreign firms such as BP or Shell to enable such overseas companies to own equity within the Russian energy sector.

The trajectory of Russia's energy system altered dramatically, however, under the reign of President Vladimir Putin, who entered into power in 2000. Putin replaced Yeltsin's *oligarchs* with his own *silovarchs*, a small cohort of elite cadres who supported "authoritarianism against liberal parties in Russia."[82] The reconsolidation of power under Putin enable Russia to reshape its foreign policy around energy exports, intended to counter the growing influence of NATO, and expand Russia's sphere of influence into Europe.[83] Much of this ascendance has been attributed to Russia's ability to use its energy resources as leverage or a "weapon" to attain higher prices or favorable agreements with countries such as Belarus, Germany, and Ukraine.

The decade 2001–10 saw three other important trends related to state ownership, exports, and electricity. First, the year 2003 saw the

Russian state move back into the petroleum sector to regain some of the control it had lost devolving equity or production agreements with foreign entities, and precipitating a forceful distribution of oil and gas property back to the state.[84] Second, in November 2009, the government adopted "Energy Strategy to 2030," a roadmap which included the target of diversifying oil and gas exports into Eastern markets, with 20–25 percent of Russia's oil exports intended to be directed at Asian markets by 2030.[85] Third, state planners initiated a program of massive reform, or restructuring, of electric utility markets.[86] As part of this reform process, generation assets were unbundled and privatized, and new investment mechanisms were integrated. A wholesale spot electricity market covering European Russia, the Urals, and Siberia was erected and economic regulation and open access arrangements for transmission networks were introduced. Cost-reflective tariffs were pursued and regulatory institutions were established,[87] likely undergirding positive performance on related metrics for electricity.

Looking ahead, one of the central dilemmas in Russian energy policy is its unique distribution of power in public-private partnerships. Russian energy strategy is dominated by the state, yet it also seeks to benefit from free markets and liberal trade.[88] Projections suggest that at least $200–360 billion of investment will be needed to modernize existing coal-fired power plants by 2030.[89] The 2030 Strategy also estimated that the oil and gas sector will require at least $2.4–2.8 trillion in investment to reach its goals.[90] Moreover, Russia has recognized the importance of renewable energy for the modernization of its economy,[91] one that could grow even further if Russia were linked with the European Union through better electricity interconnections.[92] A final future shift could be towards Arctic energy resources. Though it has the risk of becoming a "false start," the end of the previous decade did see a Russian flag-planting expedition to the North Pole and increased discussion about harnessing the hydrocarbon resources there.[93] Thus, Russia's previous, historical trends may differ markedly from its prospective trends.

Brazil

The energy crises of the 1970s had a profound and lasting impact on how the state interacted with and managed energy markets. When the first oil shock occurred, Brazil was heavily dependent on imported petroleum and saw the country's import bill rise from $600 million in 1973 to $2.5 billion in 1974, an amount equivalent to one-third of all national imports and 50 percent of currency earned from exports.[94] Brazil launched its *Proálcool* program in November 1975 to boost ethanol production and substitute ethanol for petroleum in conventional vehicles. By 1981, a mere six years later, 90 percent of all new vehicles sold in Brazil could run on ethanol – greatly minimizing dependence on imported oil.[95] The 1980s also witnessed a meaningful shift in the national management of energy markets, as Brazil transitioned from an authoritarian, state-interventionist approach to a more competitive and market-driven one. That decade, Brazil abandoned its policy of import substitution, removed protection schemes for Brazilian producers, and started the process of electricity market restructuring and liberalization.[96]

The early 1990s saw Brazilian planners continue their efforts to reduce dependence on imported oil and stimulate national self-sufficiency in energy, with policies to increase domestic oil production, expand ethanol production, generate nuclear power, and implement energy efficiency.[97] In 1993, the government continued its reform of electricity markets by ending tariff equalization for the industry, annulling some $23 billion in debts to give the sector a "clean slate," and establishing the public-private consortia SINTREL (National System of Electric Transmission) to manage access to the grid.[98] In 1997, the government passed the Brazilian Petroleum Investment Law to oversee deregulation and restructuring of the oil sector, increase natural gas use, and funnel investment into electricity supply.[99] These efforts in tandem saw domestic oil output rise from about 0.2 mb/d in the 1980s to about 1.4 mb/d in 2000, meaning domestic output met about 80 percent of Brazil's total use of oil.[100] Investments in the energy sector over the 1990s also averaged about 9 percent of total capital investments in the national economy, much of it provided by the private sector.[101]

The 2000s were punctuated by a major supply crisis in 2001. The wave of privatizations in electricity and petroleum in the 1990s meant that few upgrades in efficiency were generally made, and that electric utilities such as *Electobras* did not adequately invest in planning and capacity additions.[102] Drought conditions in late 2000 and early 2001 left numerous hydroelectric reservoirs close to empty, leading to rolling blackouts and a "power crisis" that saw a 6 percent drop in energy demand.[103] In response, the Cardoso government launched a "crash construction program" and completed 19 new natural gas power plants (with a combined capacity of 4,012 MW) to help diversify electricity supply.[104] Also, the Brazilian government started incentivizing flex-fuel vehicles (FFVs) in 2003 through reduced tax rates and fuel taxes, meaning 20 million FFVs were purchased by 2007, further lessening the strategic importance of oil.[105] The year 2004 also saw the launch of the Alternative Energy Sources Incentive Program (*Proinfa*) which centered on promoting electricity from biomass, solar photovoltaics, and wind turbines.[106]

Looking to the future, one of the "fundamental dilemmas" in Brazilian energy policy is its unique mix of public-private interventions in energy markets, namely that the government seems to still prefer a state-controlled approach but wants to rely on foreign capital and investment to meet growing infrastructural needs.[107] The recent push in the electricity sector toward natural gas, nuclear power, and other renewables has made some inroads at diversification, but in 2010 hydroelectricity still generated 80 percent of national supply and constituted 70 percent of installed capacity. Goldemberg notes that despite its potential, "energy efficiency so far has had a small role in the energy planning of Brazil."[108] Furthermore, the discovery of large domestic oil fields in 2008 adds a degree of uncertainty to future energy projections. On the one hand, tapping these fields for domestic use could increase production and self-sufficiency, arguably good things, but they could also have the effect of "locking out" renewables, efficiency, and cleaner forms of energy, as large sums of human, financial, and technological capital may be invested into fossil fuels.[109]

South Africa

South Africa is the most industrialized country in Africa, with a "highly energy intensive economy" compared to its neighbors,[110] making it a regional heavyweight in terms of economic and political influence.[111] Prior to 1994, South Africa was infamously an apartheid state, and thus energy access and development was highly uneven.

However, South Africa underwent a major social and political change in 1994, when the apartheid regime was replaced with a democratic dispensation under the leadership of Nelson Mandela. This, among other things, ended a history of economic isolationism and saw the lifting of economic sanctions, actions which created greater access to global trade of energy fuels and technology.[112] With increased trade came greater dependence on imports. For example, South Africa increased its oil imports from 55 million barrels in 1994 to 142 million barrels by 1998.[113]

Almost two-thirds of the South Africa population lacked access to the grid in 1994, meaning the new government considered electricity provision "as very important for the growth and development of the country."[114] The post-1994 government proposed and proceeded to implement a Reconstruction and Development Program which included aggressive targets for the extension of state services to be subsidized by the state, one of which was tied to the National Electrification Program, or NEP. The NEP intended to add 450,000 household connections *each year* from 1994 to 1999, setting an international precedent, and reaching about 2.5 million homes in a period of five years.[115]

The 2000s saw continued support for electrification. The NEP reached into its second and third phases, ultimately connecting 74 percent of households by the time it ended in 2007 – a jump from just 36 percent when it started in 1994.[116] Such increased access had the unfortunate side effect of also seeing per capita consumption rise, and with it, energy-related CO_2 emissions, given that the majority of the country's emissions (higher than 60 percent) come from the electricity sector, and that more than 85 percent of electricity generated by Eskom, the national provider, is coal-fired. The end of the decade was marked by a wave of massive blackouts in 2008 that crippled large portions of South Africa's economy.[117] The blackouts forced extended shutdowns of mines for weeks, with the African Development Bank estimating lost revenue from closed gold and platinum mines, responsible for 25 percent of South Africa's exports, at ZAR200 million per day.[118] As a result, planners committed to building one of the largest coal-fired power plants in the world, the 4,800 MW Medupi Power Station,[119] which went online in mid-2015.

Looking to the future, at least four separate energy security issues linger on the horizon. One is rapid exhaustion of domestic energy resources. Given such rapid rates in increases in consumption and electrification, one study projected that proven oil and gas reserves will last only 23 years before they are depleted.[120] Another is the need to diversify away from coal, which still accounted for more than 90 percent of national electricity supply in 2016.[121] Doing so would likely require the removal of subsidies for fossil fuels, a better regulatory framework for independent power producers, the provision of credit and financing for low-carbon energy systems,[122] and adhering to the stated goals of the National Development Plan (2010–15), the New Growth Path, and the National Strategy for Sustainable Development, which all have stipulations for renewable energy. Third, there is the problem of unequal access to energy even with high electrification rates. Studies have warned that South

African electrification, even when affordable, tends not to provide rural homes with everything they need. Follow-up field research in South Africa has confirmed that "substitution of traditional fuels by electricity for cooking is a slow process."[123] Other surveys of rural villages in South Africa have also documented that between 10 and 73 percent of them use batteries as their primary source of electricity, with an average across the sample of 32 percent.[124] A final issue concerns the role of energy megaprojects and interconnection of South Africa with other regional partners, in order to access their electricity or energy resources. One proposed project is for an "Africa Clean Energy Corridor," seeking to promote integration and trading of renewables across from Egypt to South Africa.[125] A final issue is integration of South African electricity markets with the 44,000 MW Grand Inga Dam being constructed in the Democratic Republic of the Congo.[126] Such projects have the potential to alter South Africa's future energy and electricity pathways considerably.

Conclusion

This overview shows how, all over the world, national governments are attempting to steer energy markets and systems through a mesh of at times conflicting, and at other times complementary, energy policies. Several conclusions can be drawn.

The first, a point we are not the first to make, is that energy policy trends are complex and at times contradictory. One set of current trends promotes environmentally sustainable economic development through notions of energy efficiency, frugality, and renewable energy resources. Another overwhelmingly depends on polluting and wasteful nonrenewable energy sources to drive economic growth. In the current (transition) phase, these are exacerbated by the challenges of overcapacity, market reform, and central-provincial dynamics. One overarching lesson is that the twin goals of sustainability, often underpinned by clean energy, and economic development, often underpinned by dirty energy, are inconsistent, and that when it comes down to it, economic development – a logic or rationale of industrial production – has often trumped environmental sustainability in importance.

Second, despite the diversity of countries and regions examined, *none* of them have been immune from some form of energy crisis, whether caused by external factors (international oil embargoes, trade disputes) or internal factors (blackouts, contention over market restructuring). Thus, as much as energy policy may seek to be proactive, it is all too often reactive, and shaped by events beyond its control.

Third, and related to this point about crisis-driven policy, is a stark lesson: domestic energy trends and geopolitical relationships can change significantly in a matter of a few decades. The United States has gone from a major energy exporter, to importer, and now back to exporter in a series of tumultuous decades. Similarly, globally, before the 1990s, China was entirely self-sufficient in its energy economy and it moved away from the influence of international energy markets, especially given the effect of several wars in the Middle East and the oil embargoes of the 1970s. Even after the 1990s, China maintained an energy supply that was 90 percent self-sufficient, but such an advantage has deteriorated sharply over the past decade. The lesson here is equally sobering: even positive energy security or reliability or environmental trends do not last forever, putting even progressive national energy policies at the whim of global markets, political shifts, and changing patterns of energy production and use.

9
Global energy governance

The energy sector raises innumerable collective action problems that states cannot solve by acting on their own. Think about transnational issues such as resource scarcity, volatile oil prices, energy poverty, nuclear proliferation, and climate change. Addressing these cross-border issues requires some form of governance at a scale beyond the nation-state. But how exactly is energy governed internationally, by whom, and with what consequences? There is no simple answer to these questions.

When oil prices soared and broke all historical records in July 2008, Mohammed ElBaradei, who was then the director of the International Atomic Energy Agency (IAEA), wrote an article in the *Financial Times* entitled: "A global agency is needed for the energy crisis." His diagnosis was as stark as it was startling:

> We have a World Health Organisation, two global food agencies, the Bretton Woods financial institutions and organisations to deal with everything from trade to civil aviation and maritime affairs. Energy, the motor of development and economic growth, is a glaring exception.[1]

At the international level, there is no global energy organization comparable to, for example, the WTO or the IMF.

Yet, to conclude that there is no global energy governance whatsoever is wrong. In fact, energy issues are discussed within a plethora of international bodies, including the United Nations, the World Bank, the IEA, and the G20. More than anything, the global energy architecture is characterized by a lack of coherence and a lack of authority, not a lack of institutions.

Since energy is so closely related to national sovereignty, states have been reluctant to transfer authority over energy issues to multilateral organizations.[2] This is even true within the European Union (EU), arguably the world's most advanced experiment with regional integration. In recent decades, the systems and networks that have developed to ensure the continuous flow of energy have become increasingly complex and transnational. As a result, states have become more integrated into international energy markets and exposed to the risks and consequences of the global energy system. The result is what McGowan has called a "paradox of sovereignty," whereby states have less control over energy policy – due to the globalization of energy markets and related externalities – but remain largely unwilling to act jointly and create strong multilateral organizations.[3]

This chapter maps the fragmented landscape of global energy governance. First, we outline the rationale and key goals of global energy governance. Second, we chart and describe the functions and evolution of key international organizations that states have set up to cooperate on energy issues. Third and finally, we take a closer look at the role that non-state actors such as companies and civil society actors can play in managing collective action problems related to energy.

The goals of global energy governance

To understand global energy governance, it is first necessary to ask: What is governance? Governance broadly refers to any of the multitudinous processes or institutions in place by which people set and enable rules needed to reach desired outcomes.[4] While most commonly envisioned as the domain of governments, many other actors are involved in governance, including civil society organizations, corporations, and institutions of finance. These can sometimes be called the "governors" of energy, those making important decisions intended to address common challenges. As we will see later in this chapter, primary and significant governors include producer and consumer clubs, economic, financial and trade organizations, and sectoral organizations.

Therefore, global energy governance can be defined as the process of making and enforcing rules to address energy-related collective action problems at a scale beyond the nation-state.[5] There are several collective action problems that arise in the energy sector, which are beyond the capacity of individual states to manage effectively.

The concept of "global public goods" helps us understand why a policy field like energy requires international governance.[6] Energy sources (such as crude oil and natural gas) and technologies (such as PV solar panels) are, of course, private goods. Yet, the energy system as a whole produces a series of outcomes that impact the wider public – it affects global public goods. Global public goods are goods that, once provided, offer benefits that are both "non-excludable" (no one can be prevented from enjoying them) and "non-rival" (anyone's enjoyment of the goods does not impinge on the consumption opportunities of others).[7] Global public goods are often under-provided, mostly because of free-riding dynamics. If no one can be excluded from the benefits of such goods, why not let others bear the brunt of supplying it? The supply of global public goods therefore depends on international cooperation.[8]

At a minimum, global energy policy should focus on ensuring the smooth operation of existing international energy markets such as the oil or gas markets. International regulation could deal with market distortions (such as monopolies, cartels or trade-distorting subsidies), facilitate and protect cross-border energy investments (to ward against instances of "resource nationalism," for instance, whereby countries block foreign investments in their energy sectors), and overcome information asymmetries (which could translate into significant price volatility, which is not conducive to energy planning and investment).[9] These functions of global energy governance already relate to a number of global public goods, such as oil price stability or a stable investment climate.

One quite fundamental problem is that markets "work" only at distributing certain types of goods. They tend to be efficient at distributing private goods such as bicycles or hamburgers – where property rights can be completely defined and protected, where owners can exclude others from access, and where property rights can be transferred or sold – but less effective at public goods such as clean air or improved energy security, which need agreed upon rules or sanctions. For all the benefits it has brought to us, our existing energy system has also enabled dreadful advances in our ability to develop weapons to kill each other, made the air in our cities unhealthy to breathe, and has begun to change our climate in a way that is irreversibly destroying the very ecosystems on which our lives depend. It has also left a large number of people deprived of modern energy services. The task of global energy governance should thus not be confined to managing existing energy markets, but to radically change them so that they no longer produce these negative consequences.[10]

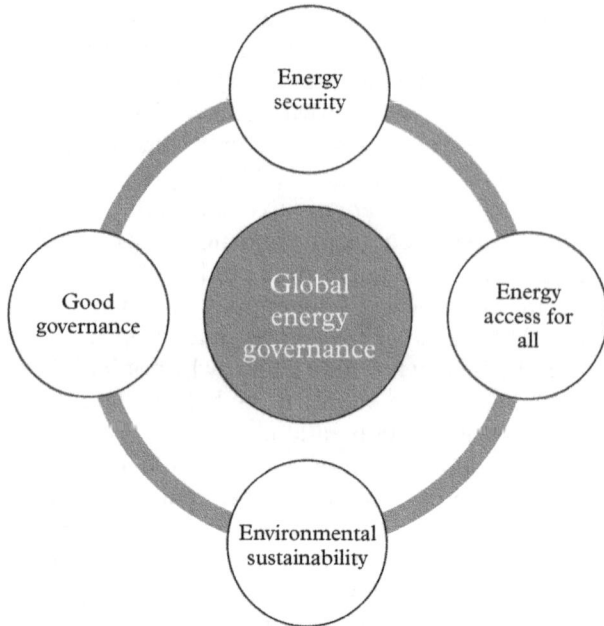

Figure 9.1   Key goals in global energy governance
Source: Authors.

Scholars have identified at least four major public goods that global energy governance should strive to bring about: energy security, energy access, environmental sustainability, and good governance (see figure 9.1).[11] Since all of these dimensions are closely connected to each other, there are actually good reasons, analytically and normatively, to consider the sustainability of the energy system as a global public good in itself.[12] Such an approach meshes well with our systems approach, which we explained in chapter 1, and is analogous to viewing the stability of the global financial system as a global public good.

First, energy security is a value that is highly prized by states, corporations, and households. It is defined by the IEA as the uninterrupted availability of energy sources at an affordable price.[13] It includes the three traditional dimensions of energy security, namely: availability, reliability, and affordability.[14] *Availability* of energy goods and services requires not just the physical resources and technologies but also the existence and proper functioning of commercial energy markets. *Reliability* involves the extent to which energy services are protected from interruption. Energy flows could be interrupted due to natural disasters, poor maintenance or underinvestment in infrastructure, terrorist attacks, or political interventions such as embargoes and sanctions. *Affordability* does not just refer to low prices for energy consumers, but also stable prices to increase planning and investment security.

Second, combating energy poverty – defined commonly as lack of access to modern energy services via electricity grids or modern cooking devices – is another major objective in global energy governance. While energy poverty may seem like a purely local or national issue, its sheer scale makes it a global issue. At present, almost 1 billion people currently have no access to electricity in their homes, depriving them of services such as lighting and refrigeration, which are essential to a decent quality of life. An even higher number of people, 2.7 billion, rely on traditional biomass for cooking, with dire consequences for their health, education, and the environment. Indeed, chapter 6 on energy justice elaborated further on some of the troubling health and environmental implications of global energy poverty. Addressing energy poverty is of critical importance to the success of the broader development agenda, yet it has long been overlooked in global governance. The inclusion of a specific goal on universal energy access by 2030 in the Sustainable Development Goals has rectified this omission, but it is clear that much greater action needs to be undertaken to close the energy access gap.

Third, our energy systems are fraught with negative environmental consequences beyond cooking and health. Some local environmental risks, such as spills, air pollution, and smog, often galvanize states into unilateral action. Other energy-related environmental problems, such as climate change or the radioactive fallout from civil nuclear accidents, do not respect national boundaries and have created demands for international governance. The challenge of climate change stands out because, if unchecked, it will create life-threatening conditions for many people on this planet. There is no way to successfully mitigate climate change without a deep and rapid decarbonization of our energy systems. International coordination has therefore focused on pushing energy efficiency and low-carbon technologies, such as renewable energy or carbon capture and storage (CCS).

Fourth, the energy sector poses problems for good governance. The upstream segment of the extractive industries has a notorious reputation when it comes to corruption, violation of labor rights, and human rights abuses.[15] As chapter 6 also notes, Shell has been accused of complicity in human rights violations in Nigeria in the 1990s, Chevron has faced lawsuits for its alleged violations of human rights of indigenous peoples in Ecuador, and Chinese oil companies have been vehemently criticized for their operations in Darfur during the genocide.[16] The transition to renewables will give a boost to the mining industries, but create similar concerns. For example, the majority of the world's cobalt today originates from the eastern part of the DRC, which has a poor record in good governance and stability.

These four key objectives can be linked to the four frames that we identified in chapter 1. Neo-mercantilists and market liberalists will emphasize the need for international cooperation on energy security, because they regard it as either a component of national security (for the former) or a key determinant of economic growth (for the latter). Environmentalists will place sustainability, and especially the fight against climate change, at the center of global energy governance. Social greens, finally, will put the eradication of energy poverty front and center, alongside the local injustices that are associated with the extractive industries.

Multilateral energy governance

The four key goals of global energy governance have not given way to a coherent international regime. There is no single, overarching global institution for governing energy. Instead, energy is governed by a Byzantine architecture of nested, parallel, and overlapping regimes, constituting what Raustiala and Victor have called a "regime complex."[17] Since energy is such a cross-cutting issue, it is hard to compile a complete list of global energy governors. Earlier attempts at mapping the global energy governance architecture have identified between 6 and 50 relevant institutions.[18] As former IEA official Neil Hirst aptly observed: "it's like gazing at the stars. The more powerful your telescope the more you will see, especially if you include bilateral as well as multilateral organisations."[19] Nevertheless, table 9.1 lists some of the key intergovernmental organizations in global energy governance. It does not mention key climate institutions, such as the UNFCCC, which have been discussed in chapter 5.

In the remainder of this section, we take a closer look at these international institutions roughly grouped into three types: first, *producer and consumer clubs* with exclusive membership such as OPEC, the IEA, and the GECF; second, *economic, financial and trade organiza-*

Table 9.1 Key institutions in global energy governance

Type	Institution	Founding	Members	Mission
Producer and consumer clubs	OPEC	1960	14	Raise oil rents for producers
	IEA	1974	30	Provide energy security for consumers
	IEF	2001	72	Global producer-consumer dialogue
	GECF	2001	12	Raise gas rents for producers
Economic, financial, and trade organizations	World Bank	1945	189	End extreme poverty and promote shared prosperity
	WTO	1995	164	Liberalize world trade in goods and services
	ECT	1994	53	Energy sector trade and investment treaty
	G20	1999	20	Premier forum for international economic cooperation
Sectoral organizations	IAEA	1957	171	Promote nuclear safety and security
	IRENA	2009	160	Promote renewable energy
	IPEEC	2009	17	Promote energy efficiency

Source: Authors.

tions, such as the World Bank, the ECT and the WTO; and, third, we focus on one *sectoral organization*, namely IRENA. Our overview begins, however, with the world's only truly universal organization, the United Nations (UN).

The United Nations

It is striking how the United Nations (UN) has long overlooked energy issues.[20] Energy was not explicitly mentioned in the UN charter. The International Atomic Energy Agency (IAEA), created in 1957, was more of a security regime than an energy regime. It did not provide advice to its member countries on their energy policies but rather only sought to ensure the peaceful, safe, and secure use of nuclear technologies. When UN specialized agencies developed activities in the energy sector these were often meant to achieve other goals, such as poverty alleviation, industrial development, or improvement of health. Even with the global conferences on sustainable development in Rio and Johannesburg (1992, 2002 and 2012), and the creation of the Commission on Sustainable Development in the early 1990s, the UN's activities in the energy sector remained fragmented and piecemeal.[21]

That situation changed in late 2015. In September 2015, members of the United Nations adopted the Sustainable Development Goals (SDGs) as the successor to the Millennium Development Goals (MDGs). Contrary to the MDGs, the SDGs contained a specific goal on energy, goal number 7, which aims to combat energy poverty and promote renewables and energy efficiency (see box 9.1). Two months after the adoption of the SDGs, the UN member states also adopted the Paris Agreement on climate change. While the text of the Paris Agreement did not mention the words "energy," "oil," "coal," "fossil

---

Box 9.1   Text of Sustainable Development Goal number 7

Goal 7. Ensure access to affordable, reliable, sustainable and modern energy for all

7.1 By 2030, ensure universal access to affordable, reliable and modern energy services
7.2 By 2030, increase substantially the share of renewable energy in the global energy mix
7.3 By 2030, double the global rate of improvement in energy efficiency
7.a By 2030, enhance international cooperation to facilitate access to clean energy research and technology, including renewable energy, energy efficiency and advanced and cleaner fossil-fuel technology, and promote investment in energy infrastructure and clean energy technology
7.b By 2030, expand infrastructure and upgrade technology for supplying modern and sustainable energy services for all in developing countries, in particular least developed countries, small island developing States, and land-locked developing countries, in accordance with their respective programmes of support.[22]

fuels," or even "carbon," the headline goal to limit global warming to "well below 2°C" basically boils down to a commitment to decarbonize the world economy over the course of this century. Even if it remains to be seen to which extent the lofty goals and aspirations of the SDGs and the Paris Agreement will actually be implemented, these two global agreements did bring some normative coherence to the fragmented global energy architecture. For the first time, the UN has charted a clear course on energy for the international community to follow in the coming decades.

The Organization of the Petroleum-Exporting Countries (OPEC)

While the UN is a general-purpose organization, there are also a number of energy-specific intergovernmental organizations, the oldest of which is the Organization of the Petroleum-Exporting Countries (OPEC). OPEC was created in 1960 by five founding members: Venezuela, Saudi Arabia, Iran, Iraq, and Kuwait. Its membership has since grown to 14 countries in 2019 (see figure 9.2), who jointly control about 40 percent of world oil production and 80 percent of reserves.[23] Needless to say, this is a diverse group and there are some powerful internal tensions such as between Iran and Saudi Arabia, which are regional rivals. In 2018, Qatar left the organization after nearly six decades of membership, due to a diplomatic feud with Saudi Arabia and the other Gulf countries over Qatar's role in the Arab Spring and its relationship with Iran. There have been other notable shifts in the membership. For instance, Ecuador and Gabon left the organization in the 1990s, but re-entered in the 2000s. Indonesia, the only Asian member country of OPEC, terminated its membership in 2009 because it had become a net oil importer a few years earlier.

OPEC's founding should be placed in the context of decolonization and the growing assertiveness of the Global South. OPEC was initially created, not as a cartel, but as a club of oil-exporting countries who wanted to stand stronger in their negotiations over royalties and tax questions with the private oil companies who then still ruled the scene. The creation of OPEC was also a reaction to the US oil import controls, enacted by President Eisenhower, which were particularly dire for Venezuela.[24] It was only 13 years after its founding, in October 1973, that OPEC wrested from the companies the power to set and change prices of their oil exports. Between October

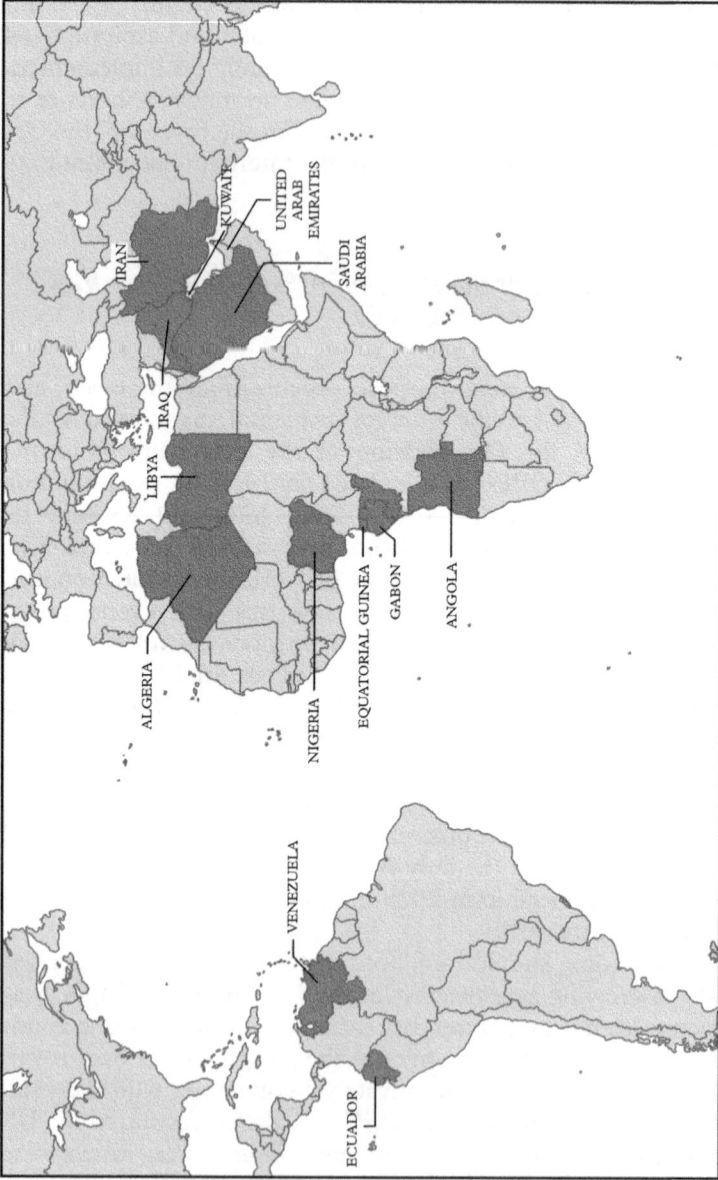

Figure 9.2 OPEC's membership, 2019
Source: Authors.

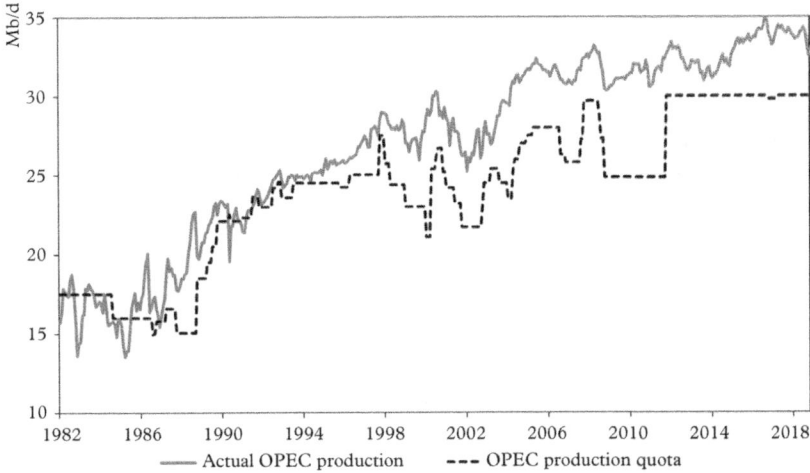

Figure 9.3 OPEC's quotas versus actual production, 1982–2019

Sources: Compiled with data from OPEC (2018) *Annual Statistical Bulletin*; EIA (2019) *April 2019 Monthly Energy Review*, accessible at https://www.eia.gov/totalenergy/data/browser/?tbl=T11.01A.

and December 1973, OPEC implemented a fourfold increase in oil prices.[25] Some OPEC countries also put into effect an embargo against the United States and its allies in the wake of the 1973 Arab–Israeli War, but this embargo was only short-lived and partially offset by increased shipments from non-Arab producers.[26] This was also the period in which many OPEC countries partially or wholly nationalized their oil industries, including Libya (1970), Algeria (1971–4), Iraq (1972), Venezuela (1974), Kuwait (1975–7), and Saudi Arabia (1973–80).

OPEC has only attempted to function as a cartel since 1982, when its members started to agree on joint production cuts. However, its effectiveness as a cartel is highly questionable. One study found that, from 1982 to 2009, OPEC countries cheated on their quota a staggering 96 percent of the time.[27] Figure 9.3 illustrates the endemic cheating. Compliance among OPEC members has been difficult to enforce because the organization lacks the means to punish members that cheat. The only actor with the power to punish cheating member states is Saudi Arabia, which is widely recognized as the informal leader of OPEC. Saudi Arabia has the world's second-biggest oil reserves (only after Venezuela), it has some of the lowest "lifting costs" (that is, the costs of getting the crude oil out of the ground), and it is the only OPEC member to command significant amounts of "spare capacity" (that is, the volume of production that can be brought on within 30 days and sustained for at least 90 days). This gives it the power to ramp up production quickly to drive other (high-cost) producers out of the market, such as it did in 1986.

Today, OPEC faces an increasingly challenging environment. In the short term, the "fracking revolution" in the US has unlocked large swaths of new oil and gas supplies, contributing to a global glut and a dramatic fall in oil prices in 2014. In the medium to long term, there is the specter of climate policies – epitomized by the adoption of the Paris Agreement in 2015 – which casts doubt on whether OPEC countries will be able to pump and monetize all of their reserves. The self-proclaimed cartel has failed to adopt a coherent, united stance in response to these colluding challenges. From 2014 to 2016, the organization failed to agree on production cuts, which triggered a battle for market share both inside and outside the cartel. As from 2016, the OPEC countries have agreed to ration production in cooperation with a group of non-member countries, including Russia, Kazakhstan, and Mexico.[28] This has been somewhat successful and prices have recovered from their low-point of less than $30 a barrel in January 2016 to more than $60 a barrel throughout most of 2018 and 2019.

The International Energy Agency (IEA)

OPEC's counterpart on the consumers' side is the International Energy Agency (IEA). The roots of the IEA lie with the Arab oil embargo of October 1973, which set off an unprecedented global energy crisis. Oil prices quadrupled almost overnight, inflation soared and the global economy began to sputter.[29] The major oil-consuming countries reacted to this crisis in an uncoordinated and competitive manner. Some pressured their oil companies into giving them preferential treatment. Others imposed restrictions on the export of petroleum. European countries sought to distance themselves from the Dutch and appease the Arabs. As Robert O. Keohane observed: "for each country in its own way, it was *sauve qui peut* [every person for himself]."[30]

In response, then US Secretary of State Henry Kissinger convened an energy conference in Washington in February 1974. Kissinger argued that the actions of OPEC could cause the "economic strangulation" of the West and constitute a casus belli if they were carried any further.[31] Kissinger did not resort to military action, however. Instead, he proposed to establish the IEA as an oil consumers' cartel to counter OPEC's influence in world oil markets. On European and Japanese insistence, the anti-OPEC stance was toned down. Japan and Europe were much more vulnerable to oil supply disruptions than the US (which, at the time, imported only one-third of its oil needs), and thus they were wary of antagonizing the very nations that were in control of their economic lifelines.[32]

By November 1974, agreement was reached among 16 Western governments on the International Energy Program, establishing the IEA. For practical purposes, the IEA was created as a daughter organization of the OECD, a decision that still has far-reaching consequences for the IEA's functioning today. Since the IEA was created at the height of a dramatic oil shock, the member states wanted the new agency to be operational as quickly as possible. Therefore, they chose to integrate the IEA into the existing institutional machinery of the OECD rather than set up a completely new, stand-alone organization from scratch. This choice meant, and still means, that only OECD member countries are eligible to join the IEA – a classic instance of "path dependency."[33]

The IEA was launched with two principal tasks. Set up as an insurance regime for the major oil consumers, the IEA's first and foremost function is to serve as an "oil market fire brigade." It does so by managing the "oil-sharing system," an emergency management scheme to cope with potential oil supply shortfalls. Every IEA member country is obliged to keep strategic oil reserves equivalent to 90 days of imports. In times of oil supply disruptions, these oil stocks could be tapped and even shared with other member countries, in accordance with a complex formula developed in the IEA's treaty and overseen by the agency's secretariat.[34] So far, the IEA has coordinated releases of strategic oil stocks only three times in its entire history. The first time was in response to the Iraqi invasion and the first Gulf War in 1990–1; the second joint stock release occurred in 2005, in the wake of Hurricanes Katrina and Rita; and the third strategic intervention by the IEA on the oil markets happened in the summer of 2011 because of the oil production fall-out caused by the Libyan civil war.

A second major role that the IEA assumes today is that of "energy watchdog": the IEA expends a great deal of staff time and resources on gathering and compiling energy statistics, facilitating peer reviews of member states' energy policies, providing energy policy advice, doing forecasts, and building scenarios. This information function has become the hallmark of the IEA's day-to-day functioning. The IEA has built up a wealth of policy wisdom on energy security matters and is now widely recognized as a leading knowledge center on

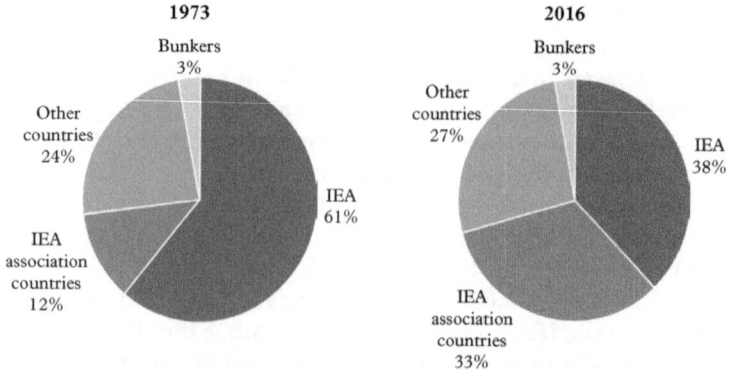

Figure 9.4   The IEA's shrinking share of world energy demand

Note: Percentages denote the regional shares in total primary energy demand (in Mtoe); bunkers include international aviation and shipping.

Source: IEA (2018) *Key World Energy Statistics.* All rights reserved; as modified by Polity Press Ltd.

energy statistics and best policy practices. The *World Energy Outlook*, the IEA's annual flagship report, has become an international point of reference in the energy industry. Some of our colleagues have even colloquially referred to it as "the energy bible."

Energy markets have changed considerably since the establishment of the IEA. New energy-consuming heavyweights have emerged outside of the OECD, most notably China and India. In 1973, IEA member countries accounted for 61 percent of global energy consumption. By 2016, their share had shrunk to about 37 percent (see figure 9.4). In most projections, this downward trend is set to continue. It is becoming ever more clear that the IEA can no longer aspire to the global role its name implies if it does not find appropriate ways to accommodate rising powers such as China and India. The rise of new powers has implications for the IEA's core functions. First, it could render the IEA's emergency oil-sharing provisions less effective. Oil demand in the IEA is expected to fall slightly in the coming decades, a drop that will be more than offset by booming oil demand from emerging economies. The net result is that the IEA's strategic oil reserves represent an ever-shrinking share of global oil consumption, muting the impact of an IEA joint oil stock release on oil markets, both in terms of volume and price. Second, it affects the IEA's access to key energy data. Without the new class of energy consumers on board, the IEA's ability to remain a leading knowledge center on energy market data is severely hampered.[35]

However, there remain strong obstacles for closer engagement with countries such as China and India. First, only OECD member countries are eligible to join the IEA. To join the OECD, applicant countries have to demonstrate that they are democratic, have market-based economies, and respect the rule of law and human rights. Some of these criteria may hinder the accession of, say, China and Russia. Second, emerging powers may have little incentive to join the OECD because they can now free ride on the IEA's oil stock regime. Third, the current voting structure of the IEA, which is based on oil consumption figures from the 1970s and thus disadvantages emerging economies, could reinforce the image that the IEA is just an undemocratic, rich men's club.[36]

At a speech on the occasion of the IEA's 35th anniversary, in 2009, Henry Kissinger, who played a leading role in the IEA's creation, said: "[The] world has changed considerably since 1973. In order to be effective in this new landscape the IEA must be prepared to evolve with it."[37] Since then, the IEA has begun to develop much closer working relations with non-member countries. The IEA has signed so-called "association agreements" with eight non-member countries: Brazil, China, India, Indonesia, Morocco, Singapore, and Thailand. These agreements – while stopping short of full membership – nevertheless allow for participation in various IEA Standing Groups and Committees as well as IEA Ministerial meetings, and have been touted as a model for global governance reform more generally.[38]

The International Energy Forum (IEF)

During the 1970s, OPEC and the IEA had a notoriously antagonistic relationship. This was illustrated during the North–South Conference of 1975–7, in Paris, which had been organized at the initiative of France in cooperation with Saudi Arabia. The conference had raised hopes for a constructive oil producer–consumer dialogue but produced disappointing results. OPEC and the IEA were unable to agree on even the most basic issues, such as continuing consultations on energy, due to rivalries at the member-state level.[39]

The result of these rivalries was that, for years, there was no direct communication between the IEA and OPEC. The two organizations communicated only indirectly, through the media. For example, in 1990, when Iraq's invasion of Kuwait took 4.3 mb/d out of the oil market, OPEC members agreed to "increase production, according to need" and called on the oil consumers "to actively participate in the stabilisation process." A few days later, the Executive Director of the IEA, Mrs Helga Steeg, issued a statement that welcomed OPEC's increase in oil production but also noted that the proposal "for a link between a production increase by OPEC and government stock draw by the IEA, as well as for a joint meeting between IEA and OPEC Ministers, is not feasible, politically or economically."[40]

Nevertheless, in the wake of the Gulf War, France and Venezuela took the initiative to organize a "Ministerial Seminar" of producers and consumers, which, with the help of the Norwegians, led to the first meeting in Paris in 1991. This meeting was very significant because, "at that time, the word 'dialogue' was seen very much as taboo."[41] Some key countries (such as the US, the UK, Japan, and Saudi Arabia) showed little interest in the dialogue and were only represented at a junior level, yet the dialogue continued throughout the 1990s in a series of biannual international energy conferences. In 2003, the process was institutionalized. A new international organization was created, the IEF, with a permanent secretariat in Riyadh. Today, the IEF has 72 member countries, including the OPEC and IEA members as well some key players from outside those blocs such as Argentina, China, India, Mexico, Russia, and South Africa.

Every two years, the IEF convenes a meeting of its members' energy ministers. The discussions center mostly on oil and gas issues. One issue in which the IEF has made progress is oil and gas data transparency. In 2001, six international organizations took the initiative to gather monthly oil statistics through a standardized questionnaire: the Joint Organizations Data Initiative (JODI). The IEF secretariat took over the coordination of this database in 2005, and the initiative was later extended from oil to gas. The database is publicly available, but the timeliness, coverage, and reliability of the data are not always up to standard. Overall, the IEF remains rather constrained since the organization lacks a firm structure or authority. It is no decision-making forum. Rather, its usefulness lies primarily in the opportunity it creates for bilateral contacts through its biannual ministerials.[42]

Gas-Exporting Countries Forum (GECF)

There have been some attempts to mimic the OPEC experience in natural gas exports. For example, Algeria tried to create a gas exporters' grouping in the early 1980s, but its effort failed due to active counter-lobbying by the US.[43] At the turn of the millennium, a new initiative by Russia, Algeria, Egypt, and Qatar proved more successful. More than a dozen gas-producing countries came together at a meeting in Tehran in 2001. The dialogue became an annual event and continued on an informal basis until, in 2008, a formal charter for the GECF was adopted. The GECF became operational as an international organization as of October 1, 2009, when five countries presented their ratification instruments: Russia, Qatar, Libya, Trinidad and Tobago, and Algeria. They have since been joined by seven additional countries, with an additional seven becoming observer countries. The 12 member countries together control more than 60 percent of the world's natural gas reserves, but just three countries – Russia, Iran, and Qatar – already account for almost 50 percent of the world's reserves (see table 9.2). Several key natural gas producers are not member of the GECF process, including the US, Australia, and Turkmenistan.

Some experts state that, for now, the GECF is not a "gas-OPEC." There are three main structural aspects of the gas market that make it difficult for the GECF to act like OPEC. First, the gas market is not truly global, because it is very expensive to transport natural gas long distances. Prices vary between regions and, as a result, cutting production in one market does not guarantee higher prices in another market. Second, the majority of natural gas is traded across borders on the basis of long-term supply contracts between producers and consumers. Third, gas can also be substituted by other fuels more easily than oil.[44] For these reasons, it has proved difficult for the diverse membership of the GECF to find common ground on coordinating policies, let alone to implement production quotas.

The World Bank

Perhaps lesser known than the other institutions in this chapter when it comes to energy governance, the World Bank has deployed activities within the energy sector since its creation in 1947, when it focused mainly on the postwar reconstruction of Europe. The World Bank is actually a group of five more distinct entities: the International Bank for Reconstruction and Development, International Development Association, International Finance Corporation, the Multilateral Investment Guarantee Agency, and the International Centre for Settlement of Investment Disputes. The World Bank's major shareholders are the United States, France, Germany, Japan, and the

Table 9.2 GECF members and key statistics

	Share in gas reserves (2017)	Share in gas production (2017)
Members		
Algeria	2.2%	2.5%
Bolivia	0.1%	3.5%
Egypt	0.9%	1.3%
Equatorial Guinea	0.7%	0.4%
Iran	17.2%	6.1%
Libya	0.7%	0.3%
Nigeria	2.7%	1.3%
Qatar	12.9%	4.8%
Russia	18.1%	17.3%
Trinidad and Tobago	0.1%	0.9%
UAE	3.1%	1.6%
Venezuela	3.3%	1.0%
TOTAL GECF	**62.0%**	**41.0%**
Observers		
Angola	0.2%	0.1%
Azerbaijan	0.7%	0.5%
Iraq	1.8%	0.3%
Kazakhstan	0.6%	0.7%
Norway	0.9%	3.3%
Oman	0.3%	0.9%
Peru	0.2%	0.4%

Note: GECF = Gas-Exporting Countries Forum.

Sources: BP (2018) *Statistical Review of World Energy*; GECF (2018) *Annual Statistical Bulletin*.

United Kingdom, and its major borrowers are Brazil, China, India, Indonesia, Mexico, and Russia.

Today, the World Bank is the world's largest multilateral development bank that provides loans and credit to developing countries to stimulate social and economic development in an attempt to alleviate poverty.[45] It defines its mission as "reducing poverty, increasing shared prosperity, and promoting sustainable development."[46] In the period 2014–18, the Bank financed energy projects in 65 countries around the world,[47] so it has a huge potential to steer the energy investments in developing countries.[48] It is a major source of financing for energy and infrastructure projects including pipelines, oil and gas fields, and power plants.

Through its financial reach, the World Bank also exerts enormous weight on macroeconomic policy advice and has used its influence to restructure entire economies more in line with its prescriptions for proper economic governance.[49] Since the mid-1980s, the World Bank's preferred model to energy sector reforms in developing economies has been liberalization and privatization.[50] Yet, the Bank's energy-lending policy has been criticized by NGOs for its poor environmental and social record.

Since the early 1990s, the World Bank's involvement in controversial large hydropower dam projects such as the Sardar Sarovar Dam in India triggered an international civil society campaign.[51] In 1997, the World Bank agreed to support an independent assessment of the development impacts of large dams through the World Commission on Dams (WCD). In its final report, which was published in November 2000, the Commission was highly critical of the development impacts of large dams and of the role of financial institutions in pushing such projects.[52]

In addition, NGOs contend that the Bank has been used as a vehicle for industrialized countries to open developing countries' fossil fuel sectors in order to satisfy the growing energy needs of the North, rather than to really address the issue of energy poverty or development.[53] In 2001, most of the World Bank's energy investments were still related to upstream activities such as electric power generation and oil and gas exploration and production, which do not necessarily increase the access of the rural poor to modern energy services.[54] In 2000, for instance, the Bank approved support for developing oil deposits in the South of Chad and for building a pipeline to the coast of Cameroon to transport the oil onto the international market.[55] It was at the time the largest single private investment in sub-Sahara Africa. However, in September 2008, the World Bank announced that it would stop supporting the project because the Government of Chad did not allocate oil revenues for poverty-reducing projects as it had promised.[56]

In response to the criticism, the Bank has introduced several major reforms, starting with the launch of an independent accountability mechanism known as the Inspection Panel in 1993. In 2013, the World Bank announced that it would "only in rare circumstances" provide financial support for new coal power stations and, in December 2017, the Bank announced that it would also no longer finance upstream oil and gas after 2019. However, in the period 2014–19, the Bank still spent US$12 billion for projects associated with fossil fuels. In fact, the Bank's energy lending for fossil fuels is three times larger than for clean energy (excluding hydro).[57]

The Energy Charter Treaty

The Energy Charter Treaty (ECT) is a comprehensive international energy treaty, created in the early 1990s, during the heyday of post-Cold War euphoria. This was a period when many Western international institutions were branching eastwards to infuse Western money and values into the transition process in the former Soviet Union. Many soviet successor states, including Russia, were rich in energy resources but needed major investments to ensure their development. Western European countries, on the other hand, saw in the region an opportunity to diversify their sources of energy supply to diminish their dependence on the Middle East. The dissolution of the Soviet Union also led to the sudden appearance of "new" transit countries along the supply routes from Russia to Europe.

The ECT itself was signed in 1994 and came into force in 1998. In principle, every country may apply for membership. The Treaty currently has a membership of 50 countries, mostly from Europe and the former Soviet Union. Substantively, the ECT covers a wide range of aspects of energy cooperation: trade, investment, transit, energy efficiency, and dispute settlement. Observers agree that the ECT has produced limited overall results in matters of trade and transit, but that its investment provisions are at its very core. The ECT's investment rules can be enforced via investor-to-state dispute settlement (ISDS) procedures, which allow foreign investors in the energy sector the right to directly sue states in international tribunals composed of three private lawyers. This has raised considerable concern that incumbent energy producers in the fossil fuel sector will use ISDS procedures to stall action on climate change.[58] The ISDS provisions effectively provide investors with rights, but not with obligations.

It is clear that the ECT process has lost some of its momentum over the years. Russia, which was arguably the most important player toward which the ECT rules were oriented, never ratified the treaty. For a while, Russia claimed to uphold the rules of the ECT on a provisional basis (that is, voluntarily, without ratifying the treaty), but it decided to withdraw from the process altogether in 2009. Russia had many grievances with the treaty, but a high-profile ISDS dispute with Yukos was probably the trigger to Russia's exit.[59] Other key players like Norway and the United States have also never been full members, hampering the ECT's reach and role in global energy governance. Even the EU, the principal driving force behind the start-up of the ECT process, seemed to lose interest in the treaty, as it began to set up new institutions, such as the European Energy Community, to create a pan-European energy market with harmonized regulatory rules. Italy withdrew from the ECT in 2016 over fears that it would face large ISDS claims from foreign investors over reforming its subsidies to solar.[60]

In recent years, a process to modernize the ECT has started. This led in 2015 to the signing of the International Energy Charter (IEC) by 75 countries. In contrast to the ECT, it is a political declaration that seeks to enhance trade and investment flows for a sustainable, secure, and affordable energy future, underpinned by the rule of law. In a sense, it is an update to the 1994 ECT but, as mentioned, it is non-binding. One notable success is that China has signed the new IEC, but it remains to be seen whether this political declaration will have any teeth.[61] The ECT, meanwhile, continues to be criticized for hampering the low-carbon energy transition, as its investment protection rules cover all of the fossil fuels, with some important renewable sources not being covered and with demand-side management categorically excluded.[62]

The World Trade Organization (WTO)

The WTO is a truly multilateral organization with a membership of about 164 countries, with another 22 countries negotiating their accession. The WTO was not specifically designed to deal with energy as a distinct sector. Yet, WTO rules are applicable to all forms of trade so, in principle, they also apply to trade in energy goods and services. This means, for instance, that energy-related goods such as crude oil or solar panels are subject to the WTO rules regarding non-discrimination, freedom of transit, and prohibition of import and export restrictions.[63]

Yet, in practice, the energy sector – which accounts for about 20 percent of global merchandise trade – has not been high on the agenda of the multilateral trade negotiations.[64] One explanation is that key energy-exporting countries such as Saudi Arabia (which joined the WTO only in 2015) and Russia (which joined in 2012) were not contracting parties of the General Agreement on Tariffs and Trade (GATT), the WTO's predecessor. In addition, GATT/WTO rules tend to focus on market access and reducing import barriers, whereas, in the energy sector, trade restrictions often take the shape of export barriers.[65] Therefore, in the first GATT negotiation rounds, there was a "gentlemen's agreement" to keep oil issues outside of GATT.[66]

In recent years, and particularly since 2010, a growing number of trade disputes concerning renewable energy technologies have firmly anchored energy issues on the WTO's agenda.[67] This rise in clean energy disputes reflects the growing scale of the renewable energy industry and the size of the market. There have been several high-profile cases. For instance, in 2010, China put in place export restrictions on rare earths, leading the US, Japan, and the EU to file a dispute against it. Japan, lodged a complaint against Canada for the Ontario province's feed-in-tariff, which included local content requirements and could thus be seen as unfairly discriminating against foreign renewable energy products. Trade friction has also led to domestic anti dumping and countervailing measures being implemented, for instance, by the US and EU against Chinese solar panels, and China on imports of polysilicon (a key input to solar PV panels) from the EU.[68]

At the heart of these disputes lies a tension between the multilateral trade rules, which are geared to counter protectionism and discrimination, and the emergence of "green industrial policies" in many countries, aimed at supporting the indigenous renewable energy industry. Handing out subsidies to the renewables sector might be trade-distorting but it could bring benefits for global decarbonization. As Dani Rodrik observes: "From a global standpoint, it would be far better if concerns about national competitiveness were to lead to a subsidy war, which expands the global supply of clean technologies, rather than a tariff war, which restricts it."[69] In 2014, a group of WTO members launched plurilateral negotiations for the establishment of an Environmental Goods Agreement (EGA), which sought to bring down tariffs and non-tariff barriers to trade in a number of key environmental products, such as wind turbines and solar panels. Since then, the number of participants has grown, with the current total representing 46 WTO members. However, negotiations have been paused since 2016 and it is currently not clear when (if ever) the negotiations will be resumed.

The G20

The G20 is often heralded as one of the most promising global institutions to act as a "steering committee" in the fragmented landscape of global energy governance.[70] Its member countries collectively represent more than 75 percent of all energy consumption and almost 80 percent of all CO_2 emissions.[71] The G20 was created as an informal club of 19 industrial countries plus the European Union in 1999, in the wake of the Asian Financial Crisis. Initially, the group met at the level of finance ministers and central bank governors, but since 2008, it has also started to come together at the level of heads of state and government. During the 2009 Pittsburgh summit, the G20 was elevated to the world's premier forum for global economic governance.

It was also in 2009 that the G20 committed to "rationalise and phase out over the medium-term inefficient fossil fuel subsidies that encourage wasteful consumption."[72] This commitment has to be understood in the context of the Great Recession, which created a need to cut public spending. It was also a maneuver by the Obama administration to create the financial and climate agenda, in the run-up to the climate conference in Copenhagen.[73] Nevertheless, implementation has been lagging and the initiation of a voluntary peer review process in the G20 has not helped much.[74] Neither has the fact that there exist differing views over the exact definition of fossil fuel subsidies.[75] Yet, the biggest reason why fossil fuel subsidies are so entrenched has to do with political economy: cutting down on subsidies will trigger protest from those who currently profit from them.[76]

The energy work of the G20 has been galvanized through the Energy Sustainability Working Group (ESWG), created under Russia's chairmanship of the G20 in 2013. This Working Group submitted its final report to the leaders of the G20 in November 2014, on the eve of the G20 Brisbane summit.[77] At that summit, the G20 leaders for the first time dedicated a full session to discuss global energy issues, thereby anchoring energy issues firmly onto the G20's agenda. In October 2015, also for the first time, the G20 organized an energy ministerial meeting in Istanbul, Turkey. Since then, the G20 leaders and energy ministers have adopted a plethora of energy and climate-related policy documents, as illustrated in table 9.3.

The International Renewable Energy Agency (IRENA)

IRENA was created in January 2009 as a stand-alone international organization dedicated to promote renewable energy. It is headquartered in Abu Dhabi, the capital of the United Arab Emirates (UAE). As of February 2019, it is arguably one of the largest intergovern-

Table 9.3 Key energy and climate-related documents adopted by the G20

Year	Presidency	Outcome document
2014	Australia	G20 Energy Efficiency Action Plan
2014	Australia	G20 Principles on Energy Collaboration
2015	Turkey	G20 Toolkit of Voluntary Options for Renewable Energy Deployment
2015	Turkey	G20 Energy Access Action Plan: Voluntary Collaboration on Energy Access
2016	China	G20 Voluntary Action Plan on Renewable Energy
2016	China	Enhancing Energy Access in Asia and the Pacific: Key Challenges and G20 Voluntary Collaboration Action Plan
2016	China	G20 Energy Efficiency Leading Programme
2017	Germany	G20 Hamburg Climate and Energy Action Plan for Growth
2019	Japan	G20 Karuizawa Innovation Action Plan on Energy Transitions and Global Environment for Sustainable Growth

Source: G20 Information Center, University of Toronto.

mental bodies dealing with energy, given it has 160 member states, with another 23 states in the process of accession. It has an annual core budget of US$21.5 million and a core staff of 70.[78] According to its Statute, the Agency's mission is to "promote the widespread and increased adoption and the sustainable use of all forms of renewable energy,"[79] including bioenergy, geothermal, hydropower, ocean energy, solar energy, and wind energy.

The creation of IRENA owes much to a diplomatic campaign spearheaded by Germany, supported by Spain and Denmark. Those three countries were dissatisfied with what they regarded as a lack of support for renewable energy technologies in established international organizations, such as the Paris-based IEA.[80] The late German social-democrat Hermann Scheer, a member of the German Bundestag from 1980 to 2010, developed a concept note to create IRENA in the 1990s and kept pushing the idea in Germany and in several international institutions. After the 2002 federal elections in Germany, the two parties in the Red-Green coalition (the SPD and the Green Party) retained a majority and the creation of IRENA was written into the coalition agreement. Those were the same political forces that pushed for the *Energiewende* or "energy transition" in Germany.[81]

From January 2009 until July 2010, when IRENA's statute entered into force, the signatory states gathered in the Preparatory

Commission, an interim body to negotiate the agency's structures and initial work program. In June 2009, during the second session of the Preparatory Commission, it was decided that the Secretariat would be located in Masdar City, the prestigious low-carbon project under construction outside of Abu Dhabi, the capital of the UAE, a significant oil exporter and member of OPEC. Once finished, the city will rely entirely on solar power and other renewable energy sources. At the same meeting, a French official, Hélène Pelosse, was elected as interim Director-General.

The choices of Masdar and Pelosse were a clear disappointment for Germany. Hermann Scheer, who pioneered the idea of IRENA, had hoped that Bonn, which already hosts the UNFCCC secretariat, would become the headquarters and that he himself would be nominated to become IRENA's first Director-General. However, the German government had refrained from nominating him because it feared his candidacy would interfere with Bonn's application to host IRENA's headquarters. This was obviously a miscalculation on the part of the German government because in the end it got neither the Secretariat nor the post of Director-General. To sugar the pill for Germany, Bonn was elected to host IRENA's Centre of Innovation and Technology. IRENA also has a small liaison office with the UN in New York. At the moment, IRENA is not officially affiliated with the UN system, although many of its internal governance structures and procedures follow conventional UN rules and regulations.

The choice of Abu Dhabi as the headquarters' location was the result of an impressive diplomatic lobby campaign that the UAE had set up, especially toward African countries. Moreover, the UAE promised to invest massively in IRENA. While Germany only offered some US$11 million, the UAE offered no less than US$136 million over the first six years. In addition, the UAE pledged to cover the entire costs for the logistics of the new agency so that the member states' contributions could be directly used for the payment of salaries and to finance the implementation of its activities. The Abu Dhabi Fund for Development also promised an amount of US$50 million each year, for the first seven years, to finance renewable energy projects in developing countries recommended by IRENA.

The first two years after the signing of IRENA's treaty were very turbulent. Less than 18 months after being elected as interim director-general, Hélène Pelosse stepped down on October 19, 2010. A number of member states were dissatisfied with what they saw as the "financial mismanagement" by Pelosse and her failure to put into practice "good accounting standards." As a result, they withheld the financial support they had promised. Among these countries, the US (which had promised about US$3 million) and Japan (which had promised close to US$2.2 million) stood out as the largest defectors. In October 2010, a new interim Director-General was appointed, Adnan Z. Amin, a seasoned UN official from Kenya. Amin served two terms and, in January 2019, the Italian Francesco La Camera was elected to the post of Director-General.

IRENA does not set standards for behavior, nor is it able to impose legally binding obligations on its members. The agency is not even designed to serve as a framework for negotiating such obligations or commitments. Instead, IRENA focuses on the gathering and dissemination of knowledge related to renewable energy technologies and policies. According to its statute, IRENA, among other things, conducts policy analysis and advice, improves technology transfer, offers capacity building, provides advice on financing, and stimulates research. While IRENA can recommend projects to be funded under the Abu Dhabi Fund for Development, a promise which was part of the bid of the UAE to host the headquarters, it does not act as a development bank.

Some concrete initiatives and outcomes of IRENA's work so far include:

- "Global Atlas for Solar and Wind," a mapping of the technical and economic potential of wind and solar energy. The Atlas database shows, among other things, the average resource conditions – annual radiation for solar mean, wind speed for wind – for all regions in the world.
- "Renewables Readiness Assessments." These are country-initiated processes that help assess the status and prospects of renewable energy deployment at the national level, identify issues that require urgent attention or promotion, and define concrete actions for the rapid upscaling of renewables.
- "Clean Energy Corridors" initiative, which focuses on promoting cross-border electricity trade and the creation of regional markets for renewable power. This initiative focuses especially on the regions of East and South Africa, West Africa, and Latin America.
- "The Small Island Developing States (SIDS) Lighthouses Initiative," which is a framework for action to support SIDS in the transformation from a predominantly fossil-based to a renewable-based and resilient energy system.

Transnational energy governance

An overview of global energy governance would not be complete without acknowledging the growing role played by hybrid or non-state actors, such as corporations, civil society actors, science organizations, subnational governments, or networks and groups that involve any multiples of the above. Efforts by these types of actors to address global issues such as energy is referred to as "transnational governance."[82] The number of international NGOs has increased consistently since World War II, and most notably since the 1990s. In 2019, a total of 5161 NGOs were registered as groups with consultative status with the UN Economic and Social Council (ECOSOC).[83]

Transnational energy governance can take many forms. Some are structured as "public-private partnerships," bringing together business actors, civil society groups, and governmental actors in joint efforts to address specific public policy problems. The UN World Summit on Sustainable Development (WSSD) in Johannesburg in 2002 was an important moment for such public-private partnerships. The Johannesburg negotiations led to so-called "Type I" outcomes, a series of legally binding intergovernmental agreements, and "Type II" outcomes, voluntary transnational partnerships in issue areas ranging from biodiversity to energy. Out of the 340 partnerships that were registered with the UN Commission on Sustainable Development (UNCSD) in early 2012, 46 had a primary focus on energy issues.[84]

The Renewable Energy and Energy Efficiency Partnership (REEEP) is a prime example of the larger universe of public-private partnerships devised and established around the 2002 summit. REEEP is a cooperative platform for more than 3,500 members and 250 registered partners, among them over 40 governmental actors (both national and subnational), including all of the G7 states, and several international organizations. The network's mission is rooted in three central areas: climate change mitigation, energy access, and encouraging energy efficiency, conservation, and demand-side management.[85]

Another instance of transnational governance is "private regimes," which involve non-state actors willing to commit to self-regulatory norms, standards, and certification schemes in a given issue area.[86] These instances of transnational energy governance are mostly voluntary in nature and tend to rely on disclosure. A prime example here is the Extractive Industries Transparency Initiative (EITI), which aims to tackle corruption in the upstream oil, gas, and mining sectors by facilitating voluntary reporting on payments made by major firms. The EITI brings together government actors, civil society groups, and corporations in the extractive industries.[87] The EITI standard is currently implemented in 52 countries around the world. Each of these countries is required to publish an annual EITI Report that discloses information: contracts and licenses, production, revenue collection, revenue allocation, and social and economic spending. The key idea is that such transparency is the best antidote against corruption and embezzlement, but the strengths and weaknesses of this approach continue to be hotly debated.[88]

Generally, there is a huge and possibly growing divergence in views about the effectiveness of transnational or polycentric energy networks. Some hail it as an opportunity. Research by Heleen de Coninck and colleagues has concluded that international technology-oriented agreements to address climate change can be effective, especially if they set standards and mandates for specific sectors, not for specific technologies.[89] Another study found that, so far, the majority of private-public energy partnerships have not been fulfilling the high expectations placed on their effectiveness. For example, by 2010, 47 percent of the energy partnerships launched at WSSD had no traceable signs of activity.[90]

Conclusion

This chapter provided an overview of global energy governance. One group of authors noted that the patchwork of institutions is often described as "chaotic, incoherent, fragmented, incomplete, illogical or inefficient."[91] The fragmentation of global energy governance can be explained on strategic, functional, and institutional grounds.[92] Strategically, the dispersion of state interests and power obstructs the construction of a comprehensive, integrated regime.[93] The interests of oil exporters and importers are often opposed to each other, which explains why they have set up their own exclusive clubs (OPEC and the IEA), rather than set up an integrated global oil regime.

From a functional perspective, the international coordination problems with regard to energy as so varied that a single institutional response is hard to organize and sustain. As Goldthau and Sovacool assert: "energy is more than a sector, policy, or field; it is instead a cross-cutting challenge that envelops a distinct set of governance challenges."[94] From an institutional standpoint, finally, there never was a "Bretton Woods" moment where a grand blueprint for global energy governance was made. Instead, many of these organizations were created in times of crisis or geopolitical shocks. The IEA, for instance, was created at the height of the 1973 oil crisis and the ECT saw the light of day after the dissolution of the Soviet Union. Often, it was deemed easier by states to create new institutions than to reform existing ones.

Clearly, the international governance of climate and energy issues has moved beyond the realms of intergovernmental processes such as the UNFCCC or the IEA, or even IRENA or a REEEP. The institutional design of the Paris Agreement, with its focus on domestically-driven and voluntary climate action, has only reinforced this trend. A "groundswell" of climate actions has now emerged as cities, regions, businesses, and civil society groups have started to step up their acts on mitigation and adaptation.[95] As a result, the climate-energy governance landscape has started to exhibit some of the characteristics of what Ostrom called "polycentric systems" – that is, systems characterized by multiple governing authorities at differing scales rather than a monocentric unit.[96] While some hail this evolution as an opportunity for broader action and policy experimentation, others are more critical of this shift. According to Bäckstrand, governance networks and partnerships:

> can lead to a "hollowing out" of the state, reinforce neoliberalism and accelerate privatization of environmental governance, [...] increased business influence, power inequalities and skewed representation of stakeholders, fragmentation of global governance, reinforcement of elite multilateralism and the retreat of state responsibility in the production of public goods.[97]

A key question for the future of global energy governance relates to coordination and orchestration of the various governance efforts in this scattered landscape. The adoption of the Paris Agreement and the SDGs in late 2015 already bring some normative coherence in that they chart key energy targets that the international community has rallied behind. Yet, we should probably expect continued calls for further strengthening, modernizing, and rationalizing of the global energy governance architecture.

10

Conclusions:
Contested energy futures

By applying four different frames to global energy politics – related to geopolitics (neo-mereantilism), the economy (market liberalism), the environment (environmentalism), and social justice (egalitarianism – we have sought to uncover the deep, and often disturbing, impact of our fossil-dominated energy system on various spheres of the modern world. The energy sector has had an enormous influence on the wealth and security of nations, our global and local environments, and – often overlooked in standard accounts of energy security – issues of social justice. All of this may come to change as tectonic shifts are currently afoot in the global energy system. For about a decade, renewable energy technologies have been coming of age and are increasingly cost competitive with conventional fuels. Their rapidly falling costs, combined with technological innovation, have set in motion the wheels of a global energy transformation.

With this in mind, we offer four synthetic conclusions arising from the book in this final chapter about the future of global energy politics. First, countries around the world will continue to be entangled in global energy interdependencies, even as this energy transformation unfolds. Fossil fuel markets will not wither away overnight and, even if some low-carbon technologies may decentralize the energy system, they are embedded in globally interdependent flows of knowledge, trade, and investment. Second, the politics of energy will remain interconnected with a host of other global issues, ranging from climate change to health to infrastructure investment. This offers opportunities for policymakers to design energy policies in ways that allow co-benefits to be reaped. Third, energy politics will remain a fulcrum for political contestation at various levels, from the global to the local. Fourth and finally, the future of energy politics is not cast in stone but is marked by considerable uncertainty.

The future politics of energy are interdependent

There used to be a time when all energy was consumed in the imme-diate vicinity of where it was produced – from firewood from a nearby forest, to local farm yields to provide food for humans and fodder for working animals. When coal started to be used in large quantities, it enabled pre-industrial societies to finally escape the "photosyn-thetic constraints" on the growth of their economies, populations, and cities – that is, their dependence on the limited amounts of energy stored in green plants and trees.[1] Tapping into the subsoil reservoirs of fossil fuels unleashed unprecedented amounts of energy. Combined with the invention of key prime movers such as steam engines and internal combustion engines, this fossil fuel revolution powered industrialization, fostered globalization, and shaped the outlook of the modern world in profound ways.

Today, as this book has revealed, energy is a truly global sector. More than 10,000 oil and gas tankers roam the oceans,[2] shipping more than 3 billion tons of oil and gas around the world every year.[3] The Turkmenistan–China gas pipeline, commissioned in 2009, makes it possible to transport gas molecules some 7,000 km eastward, all the way to Shanghai.[4] An Indian mining corporation, Adani, plans to develop one of the world's largest untapped coal reserves in Queensland, Australia, and export the coal via a new 189-km rail line to the port of Abbott Point, where the coal would be loaded onto ships destined for India and other export markets. Since 2008, a subsea electricity cable connects Norway to the Netherlands, allowing both countries to send electrons either way over a distance of 580 km.[5] An expanding network of tankers, pipelines, and electric-ity cables ties countries together in a complex web of energy trade.

This makes the politics of energy increasingly global rather than only local or national. Even individual actions taken by countries, such as the United States, or energy accidents, such as Fukushima, can have ramifications well beyond individual borders – the impli-cation being that no country is immune from the actions of others. The ongoing "shale gas boom" in the United States has motivated many coal suppliers to focus on exports rather than domestic use, meaning more coal is now being combusted in Europe and Asia.[6] Analogously, the nuclear accident in Fukushima, Japan in March

2011 led to higher global prices for liquefied natural gas (LNG) as more cargoes were diverted from Europe to Japan to make up for its shortfall in electricity.[7] Global uranium prices dropped 25 percent, and gas and coal prices increased 13.4 percent and 10.8 percent, respectively.[8] It also prompted other countries, such as Germany, to adopt plans to abandon nuclear energy, underscoring the global ramifications of a single energy-related event.

Investment is another area of global interdependence – both the management and ownership of most energy infrastructural projects are international. Roughly 40 percent of infrastructure projects around the world had capital value equal to or greater than $10 billion, meaning they required large international consortiums of financiers across numerous countries to back them.[9] The $4.6 billion Baku Tbilisi Ceyhan oil pipeline, for instance, was financed by a confederation led by BP and nine other American, Turkish, Norwegian, Italian, French, and Japanese oil companies, the World Bank Group's International Finance Corporation, the European Bank for Reconstruction and Development, the export credit agencies of seven countries, and a syndicate of 15 commercial banks. Even a single exploratory oil rig, such as the Deepwater Horizon (which famously exploded in the Gulf of Mexico in 2010), was built by Hyundai Heavy Industries in South Korea, flagged by the Marshall Islands, operated by the Swiss conglomerate Transocean, and leased to a British firm (BP) to explore for oil in the United States.[10]

It is not just fossil fuels that have given rise to global interdependencies, but low-carbon energy technologies too.[11] If you own photovoltaic (PV) solar panels, there is a 70 percent chance that they were assembled in China or Taiwan.[12] Russia acted as a key technology supplier in almost half of the nuclear power plants that have been constructed since 2000.[13] Wood pellets from forests in the southeast US are increasingly exported on diesel-powered ships to Europe, where these pellets are co-combusted with coal in a (questionable) effort to abate carbon emissions and increase the share of renewables in the energy mix.[14] Even wind turbines designed, made, and deployed in northern Europe – Denmark and Norway, for instance – have most of their steel, concrete, and other materials imported and shipped from China and South Korea, rather than European firms.[15]

Adopting a systems perspective truly reveals even the extent to which all countries are entangled in global energy interdependencies. Indeed, it is hard or even impossible to imagine a single aspect of the energy system that is not part of a global production network and value chain, or that is completely disconnected from cross-border flows of finance, technology, and ideology. Even if new renewable technologies such as solar PV or batteries may be deployed at any scale, and thus offer the prospect of "decentralizing," "digitizing," or even "democratizing" energy access, one should not forget the global political economy that is behind this technology and implies everything from cobalt mining in the DRC to the assembly of solar panels in China to the data centers of Greenland and Norway.

This global interdependence between countries has created concerns over vulnerabilities and geopolitical risks, but it has also led to the emergence of institutions to "govern" the energy sector, which are increasingly global in nature. Some are intergovernmental organizations, created and funded by national governments. Others are international NGOs, not confined to any particular country or summit process, which usually have boards and receive funding from both the public and private sector. Still others are multilateral financial institutions, regional organizations that involve two or more countries as members, and even hybrid entities including everything from transnational networks of advocacy to quasi-regulatory private bodies, global policy networks, and public-private partnerships.

The future politics of energy are interconnected

This book has not just revealed the interdependence between countries, but also the manifold interconnections between the energy system and other challenges or issue areas. Energy as an issue has no fixed boundaries. It spills over and perpetually seeps into other areas such as finance, trade, security, and development. The policy of quantitative easing that the US Federal Reserve adopted after the global financial crisis of 2007–8, for example, was a key enabler of the shale gas and oil boom in the US. The US energy revolution in turn helped drive a decline in the US current account deficit, because of the reduced need for hydrocarbon imports.[16] The cushion of US shale oil also allowed the Trump administration in 2018 to pull back from the nuclear agreement with Iran and impose sanctions. But, in doing so, it has created tensions between the US and its European allies who remain major energy importers. The ripple effects of the US shale boom are a good example of the interconnectedness of energy and other major issues.

Interconnections often develop alongside broader political, economic, and social forces. Willy Brandt's *Ostpolitik* was a key factor that facilitated the construction of a large East–West gas transmission system during the Cold War. Sanctions imposed on Russia in the wake of the 2014 Crimea crisis led it to foster new energy ties with energy-hungry economies in East Asia, particularly China. China's Belt and Road Initiative (BRI), a grand strategy to foster integration between China and a string of countries in the wider Eurasian region, could lay the groundwork of future energy connectivity and trade flows and, in doing so, influence future geopolitical spheres of influence. Concern over Iran's nuclear program have led the Trump administration to renew sanctions against Iran's oil sector, which may lead China and other countries to move away from dollar-denominated oil trade.

By far the most unifying force in global energy systems today is the challenge of climate change. There are no national boundaries in the atmosphere, meaning that a molecule of carbon dioxide emitted from chopping a tree in the Brazilian rainforest contributes as much to global warming as one emitted from the burning of dung in cooking-fires in rural India or one emanating from the chimney of a coal-fired power plant in Poland. Climate change essentially stitches together all local and national energy systems, even seemingly disparate ones. It thus contributes to increased interdependence between countries but also to interconnectedness between energy and non-energy issues. After all, to the extent that emissions from deforestation, livestock, or the thawing of the permafrost increase or decrease, the allowable "emission space" for the energy sector is affected.

Other instances of transboundary pollution abound, showcasing the interconnectedness of environmental flows. Approximately two-thirds of the anthropogenic mercury emissions generated in the United States are transported outside of its borders, whereas 40 percent of the mercury falling on the country comes from elsewhere, mostly generated in Russia and the Asia-Pacific.[17] Between 50 and 70 percent of Canada's acid rain comes from the United States, and 2–10 percent of America's pollution in this category comes from Canada.[18] Turkey, Syria, and Iran have all heavily dammed the headwaters that flow into the Tigris and Euphrates Rivers for electricity and irrigation, leaving the Shatt al Arab River without enough supply for livestock, crops, and drinking, forcing tens of thousands of Iraqi farmers to abandon their fields.[19] Forest fires in Indonesia, largely caused by land cleaning necessary to accelerate palm oil plantations that manufacture biodiesel, cause transboundary haze that threatens Malaysia and Singapore every year with increased hospital admissions and diminished rates of tourism.[20]

Whether you are a military strategist, a public health officer, or a central bank director, the energy sector will loom large in your day-to-day workings. Developments in the energy-climate sphere will have large consequences for the area that you work in and, conversely, decisions that are make in your specific field could impinge on and shape the energy sector. These linkages and interconnections are as complex as they are salient at shaping global energy systems and politics. As we will see in the next subsection, they also explain why energy politics is so contested.

The future politics of energy are contested

This book has argued that some of the biggest obstacles to switching to low-carbon technologies are not technical or economic in nature, but rather socio-technical – that is, they relate to the social, political, and regulatory aspects of particular energy technology regimes. To bring an energy innovation to market at scale requires dedicated support from governments, yet energy policymaking is often hampered by bureaucratic fragmentation, short-termism, and an inability to adequately deal with the cross-cutting nature of energy policy issues – including the synergies and potential trade-offs – because of interest group politics and regulatory capture. Energy policy often develops reactively, in response to a sudden crisis or shock, and complacency quickly returns once the worst effects of the crisis have waned.

However, which technical solutions to pursue within these transitions is incredibly polemic. To this day, proponents of nuclear power are still chasing the vision put forth in the 1950s of nuclear power becoming "too cheap to meter."[21] During the US Republican Presidential campaign of 2008, the Republican Party attempted to rally behind fossil fuels such as oil and gas, which culminated in Sarah Palin leading a room full of adults in an exalted chant of "drill, baby, drill." More recently, the right-wing Tea Party in the US has created a historical alliance with the Sierra Club, known as the Green Tea Coalition, which has been fighting with evangelical zeal for increasing the use of solar power.[22] President Trump in the United States has branded the natural gas it exports as "molecules of freedom." Each stakeholder group has their own "solution" in hand to the world's energy problems and advances it with almost religious zeal.

Conflicts over the feasibility, viability, or desirability of a given technology often boil down to different frames and how they evaluate and prioritize risks.[23] Consider two competing frames – the *market liberalists* and the *environmentalists* – and their disagreement over a technological option such as CCS. Those embracing free markets see the extreme risk in abandoning the status quo (fossil fuel dependence). To them, the riskiest course of action is to transition away from conventional energy because it will erode corporate profits and alter control over the global energy system. At risk, for them, are millions of jobs and trillions of dollars of infrastructure. Their conclusion is to support CCS. Conversely, environmentalists perceive the greatest risk to lie in continuing to support a technology that continues to devastate the environment. To them, doing nothing sires future crises, and technical fixes represent unknown risks that can never be fully predicted or controlled. They argue that applying technology at a global environmental scale is suicidal – it should be avoided as a matter of precaution. Their stance is to oppose CCS.

Nuclear power offers an exemplary example of how even shared frames can be in conflict. Egalitarians might point to the risks of not having energy at all – of societies in the developing world left literally in the dark, with the extreme injustices and health impacts of energy poverty. To these individuals, nuclear power is a necessary evil – a tool which can expand access to modern electricity networks and minimize the amount of life-endangering soot and smoke choked down by mothers and young children. Other egalitarians might oppose nuclear energy due to the risks involved with its fuel cycle (such as contamination from uranium tailings on indigenous lands), its connection to weapons of mass destruction, or the massive consequences of a serious accident (which typically affect underprivileged and/or minority populations the most). Both groups agree that the risks are huge, but view them in opposite ways with opposing courses of recommended action.

These different frames and viewpoints are mirrored in inconsistent policies or even outright institutional competition between global energy governance institutions. The World Bank and the IEA, for instance, have struggled to bring their missions (development and energy security, respectively) in line with the broader imperative of decarbonization. One could say that global energy governance is under severe strain because of two colluding trends: the breakdown of the fossil bloc that undergirded much of the post-1800 industrialization, and the simultaneous (and in some ways related) shift in global power from the West to emerging economies, particularly in Asia.

Looking at the future, for reasons such as these, energy will remain a key site for political contestation at various levels, from the local to the global. The pressures of mitigating and adapting to climate change will not wither away – indeed, one could argue that the politics of climate change will increasingly become existential.[24] Given their burgeoning energy demand, the future of energy is being shaped to a large extent in the rapidly-developing economies of Asia, and particularly China. The choices that countries like China and India make will set the pace of the global energy transition, which is likely to progress at different speeds in different countries. The energy transformation driven by increasingly cheap wind and solar power will create winners and losers across the globe, and across our four frames – geopolitics, economics, the environment, and justice. The tensions and synergies emanating from this energy system change will be at the heart of global energy politics for years to come.

The future politics of energy are uncertain

Moreover, each of these distinct frames envisions a future energy world that is strikingly different from today's – but also from each of the other frames (see table 10.1). A *neo-mercantilist frame* would pursue a global energy system comprised of rationally acting, realistic states seeking to protect their independence and maximize state power, using tools like trade agreements, military interventions, or oil stockpiles. A *market liberalist* approach would instead focus on market efficacy and the use of financial instruments to raise welfare and promote energy access. The tools of this trade would be the design and trailing of new policies, regulations, and perhaps public-private partnerships. An *environmentalist* approach would place immense constraints on fossil fuels and other polluting energy sources and seek to protect ecosystems, habitats, and the biosphere at almost any cost. The tools here might be extremely aggressive policies or perhaps instead a push to reorient modern economies on values more closely attuned to sustainability rather than capitalism. An *egalitarian* approach would prioritize informed consent and representative decision-making, as well as the preservation of human rights in the pursuit of energy fuels and services. The tools of this approach might resonate more with social justice movements as well as the use of the law to enforce regulations to hold polluters accountable. Each of the four frames also elevates a particular set of threats, sources of uncertainty, or risks, from cyberterrorism and energy subsidies to carbon footprints and energy poverty (and in doing so, lowers the importance of threats outside of each frame).

Table 10.1 Frames, objectives, and visions for global energy politics

Frame	Primary objective	Primary mechanisms	Vision for the future global energy system	Sources of focus, contestation, and uncertainty
Neo-mercantilism	Security	Governments, geopolitics	Rationally acting states which trade energy and maintain and defend their independence	Energy stockpiles and reserves, military protection of energy supply channels, cyber-terrorism
Market liberalism	Economic growth	Markets, market mechanisms	Financially prudent regimes who pursue economically optimal policies to reduce poverty and promote welfare	Global energy subsidies and subsidy reform, monopolies, state-owned energy resources or infrastructure
Environmentalism	Environmental protection	Civil society groups	Aggressively low-carbon protectors who seek to dramatically minimize and prohibit pollution and destruction	Per capita and absolute greenhouse gas emissions and carbon footprints, water scarcity, air pollution
Egalitarianism	Equity and equality	Social movements	Passionately motivated advocates who fight for respect, recognition, and the protection of basic human rights	Energy and fuel poverty, electrification, human rights abuses associated with energy production and use

Source: Authors.

While it is useful to apply these different lenses, here too, deep and growing linkages across these frames are evident. A key example is the emerging divestment campaign, a transnational campaign that tries to shift investments away from fossil fuels to bring about an energy transition. The goal of these campaigners is to ward off the threat of climate change (*environmentalist frame*) but, to the extent that this campaign is successful in shifting investments, it raises concerns over potential wasted capital and "stranded assets" (*market liberalism frame*),[25] "stranded workers" and "stranded communities" in sectors like coal mining and automotive (*egalitarianism frame*),[26] and even "stranded nations"[27] – countries saddled up with large reserves of unburnable carbon, which could even lead to regime collapse (*neo-mercantilist frame*). These different viewpoints and frames are therefore very much at the heart of recent discussion on making a "just transition."

Nonetheless, despite this uncertainty, the potential reward of making this energy transition work is immense. The clean energy revolution offers a chance to keep global warming in check at tolerable levels, redistribute global wealth to a much vaster number of actors, and reduce a key source of geopolitical tensions in the world. It can also expand access to modern energy services which can simultaneously help eradicate poverty, empower communities toward new business ventures, better equalize gender relations, and prevent the death of literally millions of women and children each year. Provided that we do not leave the energy transition on auto-pilot, or narrowly reduce its politics to a single frame, but instead actively govern, celebrate diversity, and steer it in the right direction, such a new world is surely within our grasp.

Notes

Chapter 1 Introduction: Systems, frames, and transitions

1 These examples are taken from Andrew Judge's syllabus for the course 'Global Energy Politics', 2017–2018, University of Glasgow.

2 https://www.iea.org/energyaccess/database/.

3 Smil, V. (2017) *Energy and Civilization: A History*. Cambridge, MA: MIT Press.

4 Sovacool, B. K. (2014) What are we doing here? Analyzing fifteen years of energy scholarship and proposing a social science research agenda. *Energy Research & Social Science*, 1, 1–29. Hughes, L., & Lipscy, P. Y. (2013) The politics of energy. *Annual Review of Political Science*, 16, 449–69.

5 Klare, M. (2009) *Rising Powers, Shrinking Planet: The New Geopolitics of Energy*. Basingstoke: Macmillan. Högselius, P. (2019) *Energy and Geopolitics*. New York: Routledge.

6 Sovacool, B. K. (Ed.). (2010) *The Routledge Handbook of Energy Security*. New York: Routledge. Dyer, H., & Trombetta, M. J. (Eds.) (2013) *International Handbook of Energy Security*. Cheltenham: Edward Elgar Publishing. Dannreuther, R. (2017) *Energy Security*. Chichester: John Wiley & Sons. Cherp, A., & Jewell, J. (2014) The concept of energy security: Beyond the four As. *Energy Policy*, 75, 415–21. Nyman, J. (2018) *The Energy Security Paradox: Rethinking Energy (In)security in the United States and China*. New York: Oxford University Press.

7 Goldthau, A. (Ed.) (2016) *The Handbook of Global Energy Policy*. Chichester: John Wiley & Sons.

8 Van de Graaf, T. (2013) *The Politics and Institutions of Global Energy Governance*. New York: Palgrave Macmillan. Goldthau, A., & Witte, J. M. (Eds.) (2010) *Global Energy Governance: The New Rules of the Game*. Washington, DC: Brookings Institution Press. Florini, A., & Sovacool, B. K. (2009) Who governs energy? The challenges facing global energy governance. *Energy Policy*, 37(12), 5239–48. Van de

Graaf, T., & Colgan, J. (2016) Global energy governance: A review and research agenda. *Palgrave Communications*, 2(1), 15047.

9   Hancock, K. J., & Vivoda, V. (2014) International political economy: A field born of the OPEC crisis returns to its energy roots. *Energy Research & Social Science*, 1, 206–16. Van de Graaf, T., Sovacool, B. K., Ghosh, A., Kern, F., & Klare, M. T. (Eds.) (2016) *The Palgrave Handbook of the International Political Economy of Energy*. New York: Palgrave Macmillan. Goldthau, A., & Keating, M. F. (Eds.) (2018) *Handbook of the International Political Economy of Energy and Natural Resources*. Cheltenham: Edward Elgar Publishing. Kuzemko, C., Lawrence, A., & Watson, M. (2019) New directions in the international political economy of energy. *Review of International Political Economy*, 26, 1–24.

10   York, R. & Bell, S. E. (2019) Energy transitions or additions? Why a transition from fossil fuels requires more than the growth of renewable energy. *Energy Research & Social Science*, 51, 40–3.

11   Shove, E. (2003) *Comfort, Cleanliness, and Convenience: The Social Organization of Normality*. Oxford: Berg.

12   Kander, A., Malanima, P., & Warde, P. (2015) *Power to the People: Energy in Europe over the Last Five Centuries*. Princeton, NJ: Princeton University Press.

13   Morris, I. (2015) *Foragers, Farmers and Fossil Fuels: How Human Values Evolve*. Princeton, NJ: Princeton University Press. Mitchell, T. (2011) *Carbon Democracy: Political Power in the Age of Oil*. London: Verso Books.

14   Steffen, W., Broadgate, W., Deutsch, L., Gaffney, O., & Ludwig, C. (2015) The trajectory of the Anthropocene: The great acceleration. *The Anthropocene Review*, 2(1), 81–98.

15   Rockström, J., Steffen, W. L., Noone, K., et al. (2009) Planetary boundaries: Exploring the safe operating space for humanity. *Ecology and Society*, 14(2), 32.

16   US Energy Information Administration, *China*, available at https://www.eia.gov/beta/international/country.php?iso=CHN.

17   'China creates over 65 mln jobs in five years: official,' *Xinhua*, December 28, 2017.

18   Schneider, K., Turner, J. L., Jaffe, A., & Ivanova, N. (2011) Choke point China: Confronting water scarcity and energy demand in the world's largest country. *Vermont Journal of Environmental Law*, 12, 714–34.

19   IRENA (2018) *Renewable Power Generation Costs in 2017*. Abu Dhabi: IRENA, p. 3.

20   Goldthau, A., & Sovacool, B. K. (2012) The uniqueness of the energy security, justice, and governance problem. *Energy Policy*, 41, 232–40.

21   Patterson, W. (2000) Energy 21: Making the World Work. Melchett Lecture, London Planetarium, June 22.

22   Smil, V. (2006) *Energy: A Beginner's Guide*. Oxford: Oneworld Publications.

23 Sovacool, B., & Dworkin, M. (2014) *Global Energy Justice: Problems, Principles and Practices*. Cambridge: Cambridge University Press.

24 Smil, above n. 22.

25 Note that there are two other laws of thermodynamics which we do not cover here, as they are less relevant for our book.

26 Rifkin, J., & Howard, T. (1980) *Entropy: A New World View*. New York: Viking Press.

27 Soddy, F. (1911) *Matter and Energy*. London: Oxford University Press, pp. 10–11.

28 An easy tool to convert units of energy can be found at: https://www. iea.org/statistics/resources/unitconverter/.

29 MacKay, D. (2008) *Sustainable Energy – Without the Hot Air*. Cambridge: UIT Cambridge.

30 Sovacool & Dworkin, above, n. 23.

31 Example drawn from: MacKay, above, n. 29.

32 Hughes, T. P. (2004) *Human Built World: How to Think About Technology and Culture*. Chicago, IL: University of Chicago Press. Hughes, T. P., & Hughes, A. C. (2000) Introduction. In Hughes, A. C., & Hughes, T. P. (Eds.) *Systems, Experts, and Computers: The Systems Approach in Management and Engineering, World War II and After*. London: MIT Press, pp. 1–26.

33 Anarow, B., Greener, C., Gupta, V., Kinsley, M., Henderson, J., Page, C., & Parrot, K. (2003) Whole-Systems Framework for Sustainable Consumption and Production. Environmental Project No. 807. Denmark: Danish Ministry of the Environment.

34 Stansinoupolos, P., Smith, M. H., Hargroves, K., & Desha, C. (2013) *Whole System Design: An Integrated Approach to Sustainable Engineering*. New York: Routledge.

35 Definition available at its website: https://www.iea.org/topics/energy security/.

36 Patterson, W. (2009) Managing energy data. Energy, Environment and Resource Governance Working Paper. Chatham House, May, p. 26.

37 Goffman, E. (1974) *Frame Analysis: An Essay on the Organization of Experience*. Cambridge, MA: Harvard University Press.

38 Lakoff G. (2004) *Don't Think of an Elephant! Know Your Values and Frame the Debate*. White River Junction, VT: Chelsea Green Publishing, p. xv.

39 Benford, R. D., & and Snow, D. A. (2000) Framing processes and social movements: An overview and assessment. *Annual Review of Sociology*, 26, 611–39. Quote is at 622.

40 Nyberg, D., Wright, C., & Kirk, J. (2017) Re-producing a neoliberal political regime: Competing justifications and dominance in disputing fracking. *Research in the Sociology of Organizations*, 52, 143–71.

41 Nyberg, D., Wright, C., & Kirk, J. (2018) Fracking the future: The temporal portability of frames in political contests. *Organization Studies*, https://doi.org/10.1177/0170840618814568.

Chapter 2   The history and functioning of energy markets

1 Own calculations. Based on IEA (2018) *Key World Energy Statistics*. Paris: IEA/OECD, p. 22.
2 Nordhaus, W. D. (2009) The economics of an integrated world oil market. International Energy Workshop, Venice, June.
3 York, R. (2017) Why petroleum did not save the whales. *Socius*, 3, 1–13.
4 Maugeri, L. (2006) *The Age of Oil: The Mythology, History, and Future of the World's Most Controversial Resource*. Westport, CT: Praeger, p. 5.
5 Daintith, T. (2010) *Finders Keepers? How the Law of Capture Shaped the World Oil Industry*. Abingdon: Routledge.
6 Maugeri, above, n. 4, p. 19.
7 Yergin, D. (1991) *The Prize: The Epic Quest for Oil, Money & Power*. New York: Simon and Schuster.
8 Venn, F. M. (1986) *Oil Diplomacy in the Twentieth Century*. Basingstoke: Palgrave Macmillan. Toprani, A. (2019) *Oil and the Great Powers: Britain and Germany, 1914 to 1945*. Oxford: Oxford University Press.
9 Stevens, P. (2013) History of the international oil industry. In *Global Resources*. London: Palgrave Macmillan, pp. 13–32.
10 Blair, J. M. (1976) *The Control of Oil*. New York: Pantheon Books. Yergin, above, n. 7, p. 184.
11 Sampson, A. (1975) *The Seven Sisters: The Great Oil Companies and the World They Shaped*. New York: Viking Press, p. 77.
12 Markus, U. (2014) *Oil and Gas: The Business and Politics of Energy*. London: Palgrave.
13 Cowhey, P. F. (1985) *The Problems of Plenty: Energy Policy and International Politics*. Berkeley, CA: University of California Press.
14 Sampson, above, n. 13.
15 Ibid., p. 74.
16 Hartshorn, J. E. (1993) *Oil Trade: Politics and Prospects*. Cambridge: Cambridge University Press.
17 Fattouh, B., & Sen, A. (2016) The past, present, and future role of OPEC. In Van de Graaf, T. et al. (Eds.) *The Palgrave Handbook of the International Political Economy of Energy*. London: Palgrave Macmillan, pp. 73–94. Markus, U. (2016) The international oil and gas pricing regimes. In Van de Graaf, T. et al. (Eds) *The Palgrave Handbook of the International Political Economy of Energy*. London: Palgrave Macmillan, pp. 225–46.
18 Maugeri, above, n. 4, p. 47.
19 Skeet, I. (1991) *Opec: Twenty-Five Years of Prices and Politics*. Cambridge: Cambridge University Press, p. 17.
20 Parra, F. (2004) *Oil Politics: A Modern History of Petroleum*. London: IB Tauris.
21 Kapstein, E. B. (1990) *The Insecure Alliance: Energy Crises and Western Politics Since 1944*. Oxford: Oxford University Press.

22 Murkowski, L., McCain, J., & Corker, B. (2015) The US Needs to End Its Ban on Crude Oil Exports. *Foreign Policy*, April 14.
23 Yergin, above, n. 7, p. 492. McDermott, R. (1998) *Risk-Taking in International Politics: Prospect Theory in American Foreign Policy*. Ann Arbor, MI: University of Michigan Press. Kapstein, above, n. 21, pp. 109–10.
24 Yergin, above, n. 7, pp. 499–500.
25 Fattouh and Sen, above, n. 19. Morse, E. L. (1999) A new political economy of oil? *Journal of International Affairs*, 53(1), 1–29.
26 Yergin, above, n. 7, p. 523. Parra, above, n. 20.
27 Fattouh and Sen, above, n. 17, p. 75.
28 Maugeri, above, n. 4.
29 Morse, above, n. 25.
30 Yergin, above, n. 7.
31 Scott, R. (1977) Innovation in international organization: The International Energy Agency. *Hastings International and Comparative Law Review*, 1, 1–56.
32 Katz, J. (1981) The International Energy Agency: Processes and prospects in an age of energy interdependence. *Studies in Comparative International Development*, 16, 67–85.
33 Basosi, D., Garavini, G., & Trentin, M. (Eds.) (2018) *Counter-shock: The Oil Counter-Revolution of the 1980s*. London: Bloomsbury Publishing.
34 Al-Moneef, M. (2018) Saudi Arabia and the counter-shock of 1986. In Basosi, D., Garavini, G., & Trentin, M. (Eds.) *Counter-shock: The Oil Counter-Revolution of the 1980s*. London: Bloomsbury Publishing, p. 103.
35 Gaidar, Y. (2007) The Soviet collapse: Grain and oil. American Enterprise Institute for Public Policy Research. Available at https://msuweb.montclair.edu/~lebelp/GaidarAEISovietCollapseApril2007.pdf.
36 Bull-Berg, H. J. (1987) *American International Oil Policy: Causal Factors and Effect*. New York: St. Martin's Press.
37 Ikenberry, G. J. (1988) Market solutions for state problems: The international and domestic politics of American oil decontrol. *International Organization*, 42(1), 151–77.
38 Noël, P. (2002) *Production d'un ordre pétrolier libéral: une politique normative américaine dans les relations internationales entre 1980 et 2000*. Doctoral dissertation, University of Grenoble.
39 Leverett, F. (2010) Consuming energy: Rising powers, the International Energy Agency, and the global energy architecture. In Alexandroff, A. S., & Cooper, A. F. (Eds.) *Rising States, Rising Institutions: Challenges for Global Governance*. Washington, DC: Brookings Institution Press, p. 246.
40 Maugeri, above, n. 4, pp. 137–8.

41 Leverett, above, n. 39.

42 Maugeri, above, n. 4, p. 137.

43 Kleveman, L. (2003) *The New Great Game: Blood and Oil in Central Asia*. New York: Grove Press.

44 Yergin, D. (2011) *The Quest: Energy, Security, and the Remaking of the Modern World*. New York: Penguin.

45 Ibid., p. 87.

46 Claes, D. H. (2001) *The Politics of Oil-Producer Cooperation*. Boulder, CO: Westview Press.

47 McNally, R., & Levi, M. (2011) A crude predicament: The era of volatile oil prices. *Foreign Affairs*, 90(4), 100–111.

48 Yergin, above, n. 44, pp. 128–9.

49 Ibid., p. 162.

50 Ibid., pp. 165–9.

51 Hamilton, J. D. (2009) *Causes and Consequences of the Oil Shock of 2007–08*. NBER Working Paper No. w15002. National Bureau of Economic Research.

52 Rühl, C. (2010) Global energy after the crisis: Prospects and priorities. *Foreign Affairs*, 89(2), 63–75.

53 Van de Graaf, T. (2013) The "oil weapon" reversed? Sanctions against Iran and US-EU structural power. *Middle East Policy*, 20(3), 145–63.

54 "Markets 'on edge' over 'eerie calm' in oil prices," *CNBC*, June 16, 2014. Available at: https://www.cnbc.com/2014/06/16/markets-on-edge-over-eerie-calm-in-oil-prices.html.

55 Van de Graaf, T., & Bradshaw, M. (2018) Stranded wealth: Rethinking the politics of oil in an age of abundance. *International Affairs*, 94(6), 1309–1328.

56 Van de Graaf, T. (2017) Is OPEC dead? Oil exporters, the Paris agreement and the transition to a post-carbon world. *Energy Research & Social Science*, 23, 182–8.

57 Peebles, M. W. (1980) *Evolution of the Gas Industry*. London: Macmillan.

58 Victor, D. G., Jaffe, A. M., & Hayes, M. H. (Eds.) (2006) *Natural Gas and Geopolitics: From 1970 to 2040*. Cambridge: Cambridge University Press, p. 6.

59 Smil, V. (2015) *Natural Gas: Fuel for the 21st Century*. Chichester: John Wiley & Sons.

60 Victor et al., above, n. 58, p. 7.

61 Cragg, C. (2013) History of the gas industry. In Dannreuther, R., & Ostrowski, W. (Eds.) *Global Resources*. London: Palgrave Macmillan, pp. 59–76.

62 Peebles, above, n. 57, pp. 153–5.

63 Smil, above, n. 59, p. 30.

64 BP (2019) *Statistical Review of World Energy*.

65 Smil, above, n. 59.

66 Grigas, A. (2017) *The New Geopolitics of Natural Gas*. Cambridge, MA: Harvard University Press, p. 103.

67 Kandiyoti, R. (2015) *Powering Europe: Russia, Ukraine, and the Energy Squeeze*. New York: Palgrave Macmillan.

68 Högselius, P. (2012) *Red Gas: Russia and the Origins of European Energy Dependence*. New York: Palgrave Macmillan.

69 Ibid.

70 Ibid.

71 Gustafson, T. (1985) *Soviet Negotiating Strategy: The East–West Gas Pipeline Deal, 1980–1984*. Santa Monica, CA: The Rand Corporation.

72 Nies, S. (2008) *Oil and Gas Delivery to Europe: An Overview of Existing and Planned Infrastructures*. Paris: IFRI.

73 Kandiyoti, above, n. 67.

74 Jensen, J. T. (2004) *The Development of a Global LNG Market: Is it Likely? If so, When?* Oxford: Oxford Institute for Energy Studies.

75 International Gas Union (2019) *World LNG Report – 2019 edition*. Barcelona: IGU.

76 Howarth, R. W., Santoro, R., & Ingraffea, A. (2011) Methane and the greenhouse-gas footprint of natural gas from shale formations. *Climatic Change*, 106(4), 679.

77 Energy Information Administration (2018) US liquefied natural gas export capacity to more than double by the end of 2019. December 10, 2018. Available at https://www.eia.gov/todayinenergy/detail.php?id=37732.

78 Energy Information Administration (2015) World shale resource assessments. September 4, 2015. Available at https://www.eia.gov/analysis/studies/worldshalegas/.

79 Van de Graaf, T., Haesebrouck, T., & Debaere, P. (2018) Fractured politics? The comparative regulation of shale gas in Europe. *Journal of European Public Policy*, 25(9), 1276–93.

80 Williams, L., Macnaghten, P., Davies, R., & Curtis, S. (2017) Framing "fracking": Exploring public perceptions of hydraulic fracturing in the United Kingdom. *Public Understanding of Science*, 26(1), 89–104.

81 Cragg, above, n. 61, p. 41.

82 Grigas, above, n. 66, p. 49.

83 Ibid., p. 52.

84 IEA (2019) Global Energy & CO2 Status Report, March 26. Available at https://www.iea.org/geco/.

85 Thurber, M. (2019) *Coal*. Cambridge: Polity, p. 3.

86 Podobnik, B. (2006) *Global Energy Shifts: Fostering Sustainability in a Turbulent Age*. New Delhi: TERI Press.

87 Thurber, above, n. 85, p. 23.

88 Ibid., p. 29.

89 McElroy, M. B. (2010) *Energy: Perspectives, Problems, and Prospects*. Oxford: Oxford University Press, p. 114.

90 'How China cut its air pollution', *Economist*, January 15, 2018.

91 Powering Past Coal Alliance website: https://poweringpastcoal.org/about.

92 IEA (2018) *World Energy Outlook 2018*. Paris: IEA/OECD, p. 292.

93 World Nuclear Association, Plans for new reactors worldwide, October 2018, available at http://www.world-nuclear.org/information-library/current-and-future-generation/plans-for-new-reactors-worldwide.aspx.

94 IEA, above, n. 92.

95 World Nuclear Association (2019) How uranium ore is made into nuclear fuel. Available at http://www.world-nuclear.org/nuclear-basics/how-is-uranium-ore-made-into-nuclear-fuel.aspx.

96 Brutschin, E., & Jewell, J. (2018) International political economy of nuclear energy. In: Goldthau, A., Keating, M. F., & Kuzemko, C. (Eds.) *Handbook of International Political Economy of Energy and Natural Resources*. Chelentham: Edward Elgar, pp. 322–41.

97 Ibid.

98 Jewell, J., Vetier, M., & Garcia-Cabrera, D. (2019) The international technological nuclear cooperation landscape: A new dataset and network analysis. *Energy Policy*, 128, 838–52.

99 Saha, S. (2017) Russia's nuclear diplomacy. *Foreign Affairs*, April 2. Gallucci, N., & Shellenberger, M. (2017) Will the West let Russia dominate the nuclear market? *Foreign Affairs*, August 3.

100 IRENA (2019) *Renewable Capacity Highlights*, March 31. Abu Dhabi: IRENA.

101 IRENA (2019) *Renewable Power Generation Costs in 2018*. Abu Dhabi: IRENA.

102 Global Commission on the Geopolitics of Energy Transformation (2019) *A New World: The Geopolitics of the Energy Transformation*. Abu Dhabi: IRENA.

103 See https://www.irena.org/cleanenergycorridors.

104 See website of the Global Energy Interconnection Development and Cooperation Organization: http://www.geidco.org/.

105 UNCTAD (2014) *The State of the Biofuels Market: Regulatory, Trade and Development Perspectives*. Geneva: UNCTAD.

106 World Energy Council Germany (2018) *International aspects of a power-to-X roadmap*, October 18. Available at https://www.frontier-economics.com/media/2642/frontier-int-ptx-roadmap-stc-12-10-18-final-report.pdf.

Chapter 3   Energy and security: Fueling geopolitics and war?

1 Bromley, S. (1991) *American Hegemony and World Oil: The Industry, the State System and the World Economy*. University Park, PA: Penn State Press.

2 IRENA (2019) *A New World: The Geopolitics of the Energy Transformation.* Abu Dhabi: IRENA.

3 Spiro, D. E. (1999) *The Hidden Hand of American Hegemony: Petrodollar Recycling and International Markets.* Ithaca, NY: Cornell University Press.

4 Taylor, I. (2006) China's oil diplomacy in Africa. *International Affairs,* 82(5), 937–59.

5 Månsson, A. (2014) Energy, conflict and war: Towards a conceptual framework. *Energy Research & Social Science,* 4, 106–16.

6 Leonard, M. (Ed.) (2016) *Connectivity Wars: Why Migration, Finance and Trade Are the Geo-Economic Battlegrounds of the Future.* Berlin: European Council on Foreign Relations (ECFR).

7 Blackwill, R. D., & Harris, J. M. (2016) *War by Other Means: Geoeconomics and Statecraft.* Cambridge, MA: Harvard University Press.

8 Högselius, P. (2019) *Energy and Geopolitics.* New York: Routledge.

9 Ibid., p. 132.

10 Tsurumi, Y. (1976) Japan. In: Vernon, R. (Ed.) *The Oil Crisis.* New York: W.W. Norton & Company, p. 124.

11 O'Leary, N. (2014) G7 to begin reducing Russian energy dependency – Ed Davey, *Reuters,* May 6.

12 Stegen, K. S. (2011) Deconstructing the "energy weapon": Russia's threat to Europe as case study. *Energy Policy,* 39(10), 6505–13.

13 Henderson, J. (2016) Does Russia have a potent gas weapon? In Van de Graaf, T. et al. (Eds.) *The Palgrave Handbook of the International Political Economy of Energy.* London: Palgrave Macmillan, pp. 461–86.

14 Omonbude, E. (2016) *Cross-Border Oil and Gas Pipelines and the Role of the Transit Country: Economics, Challenges and Solutions.* Basingstoke: Palgrave Macmillan.

15 Stevens, P. (2009) *Transit Troubles: Pipelines as a Source of Conflict.* London: Chatham House.

16 Victor, D. G., Jaffe, A. M., & Hayes, M. H. (Eds.) (2006) *Natural Gas and Geopolitics: From 1970 to 2040.* Cambridge: Cambridge University Press, pp. 337–8.

17 Fattahi, M., & Batrawy, A. (2019) Iran seizes third foreign tanker, says it was carrying "smuggled fuel." *The Times of Israel,* August 4, available at https://www.timesofisrael.com/iran-seizes-third-foreign-tanker-says-it-was-carrying-smuggled-fuel/.

18 Navias, M. S., & Hooton, E. R. (1996) *Tanker Wars: The Assault on Merchant Shipping during the Iran-Iraq Conflict, 1980–1988* (Vol. 6). London: I.B. Tauris.

19 Talmadge, C. (2008) Closing time: Assessing the Iranian threat to the Strait of Hormuz. *International Security,* 33(1), 82–117.

20 Hughes, L., & Long, A. (2015) Is there an oil weapon? Security implications of changes in the structure of the international oil market. *International Security,* 39(3), 152–89.

21 Rodenburg, J., & Hengeveld, R. (1995) *Embargo: Apartheid's Oil Secrets Revealed*. Amsterdam: Amsterdam University Press. Crawford, N. C. (1999) Oil sanctions against apartheid. In Crawford, N. C., & Klotz, A. (Eds.) *How Sanctions Work*. London: Palgrave Macmillan, pp. 103–26.

22 Pascual, C. (2015) The new geopolitics of energy. Center on Global Energy Policy, Columbia University. Available at https://energypol icy.columbia.edu/sites/default/files/The%20New%20Geopolitics%20 of%20Energy_September%202015.pdf.

23 Torbat, A. E. (2005) Impacts of US trade and financial sanctions on Iran. *World Economy*, 28(3), 407–34.

24 Lektzian, D., & Biglaiser, G. (2013) Investment, opportunity, and risk: Do US sanctions deter or encourage global investment? *International Studies Quarterly*, 57(1), 65–78.

25 Kozhanov, N. A. (2011) US economic sanctions against Iran: Undermined by external factors. *Middle East Policy*, 18(3), 144–60.

26 Van de Graaf, T. (2013) The "oil weapon" reversed? Sanctions against Iran and US–EU structural power. *Middle East Policy*, 20(3), 145–63.

27 Högselius, above, n. 8, p. 134.

28 United Nations Security Council Resolution 2371 (2017) Adopted by the Security Council at its 8019th meeting, on August 5, 2017. S/RES/2371.

29 Wilson, J. D. (2018) Whatever happened to the rare earths weapon? Critical materials and international security in Asia. *Asian Security*, 14(3), 358–73.

30 Fattouh, B. (2011) *An Anatomy of the Crude Oil Pricing System*. Oxford: Oxford Institute for Energy Studies.

31 Crane, K., Goldthau, A., Toman, M., Light, T., & Johnson, S. E. (2009) *Imported Oil and US National Security*. Santa Monica, CA: Rand Corporation, p. 37.

32 Clayton, B., & Levi, M. (2012) The surprising sources of oil's influence. *Survival*, 54(6), 107–22.

33 Cederlöf, G., & Kingsbury, D. V. (2019) On PetroCaribe: Petropolitics, energopower, and post-neoliberal development in the Caribbean energy region. *Political Geography*, 72, 124–33.

34 Orttung, R. W., & Overland, I. (2011) A limited toolbox: Explaining the constraints on Russia's foreign energy policy. *Journal of Eurasian Studies*, 2(1), 74–85.

35 Newnham, R. (2011) Oil, carrots, and sticks: Russia's energy resources as a foreign policy tool. *Journal of Eurasian Studies*, 2(2), 134–43.

36 Henderson, J., Pirani, S., & Yafimava, K. (2017) CIS gas pricing: Towards European netback? In Stern, J. (Ed.) *The Pricing of Internationally Traded Gas*. Oxford: Oxford University Press, pp. 178–223.

37 Svoboda, K. (2011) Business as usual? Gazprom's pricing policy toward the Commonwealth of Independent States. *Problems of Post-Communism*, 58(6), 21–35.

38 Stern, J. P. (2005) *The Future of Russian Gas and Gazprom*. Oxford: Oxford University Press.
39 Pirani, S. (2014) *Ukraine's Imports of Russian Gas: How a Deal Might Be Reached*. Oxford: Oxford Institute for Energy Studies.
40 Ibid.
41 Birnbaum, M., & Carol Morello, C. (2014) Russia's Gazprom cuts gas to Ukraine in a new phase of the nation's conflict. *Washington Post*, June 16.
42 Overland, I. (2017) The hunter becomes the hunted: Gazprom encounters EU regulation. In Andersen, S. S., Goldthau, A., & Sitter, N. (Eds.) *Energy Union*. London: Palgrave Macmillan, pp. 115–30.
43 European Commission (2018) Antitrust: Commission imposes binding obligations on Gazprom to enable free flow of gas at competitive prices in Central and Eastern European gas markets. Press release, May 24. Available at http://europa.eu/rapid/press-release_IP-18-3921_en.htm.
44 Overland, above, n. 42.
45 Shaffer, B. (2005) From pipedream to pipeline: A Caspian success story. *Current History*, 104(684), 343–6.
46 Baev, P. K., & Øverland, I. (2010) The South Stream versus Nabucco pipeline race: Geopolitical and economic (ir)rationales and political stakes in mega-projects. *International Affairs*, 86(5), 1075–90.
47 Bazilian, M., Sovacool, B., & Moss, T. (2017) Rethinking energy statecraft: United States foreign policy and the changing geopolitics of energy. *Global Policy*, 8(3), 422–5.
48 Högselius, above, n. 8, p. 140.
49 Kong, B., & Gallagher, K. P. (2017) Globalizing Chinese energy finance: The role of policy banks. *Journal of Contemporary China*, 26(108), 834–51.
50 IRENA (2019) *A New World: The Geopolitics of the Energy Transformation*. Abu Dhabi: IRENA.
51 Ibid.
52 Small, A. (2018) The backlash to Belt and Road. *Foreign Affairs*, February 16.
53 Hurley, J., Morris, S., & Portelance, G. (2018) Examining the debt implications of the Belt and Road Initiative from a policy perspective. CGD Policy Paper, Center for Global Development.
54 Gallagher, K. P., Kamal, R., Jin, J., Chen, Y., & Ma, X. (2018) Energizing development finance? The benefits and risks of China's development finance in the global energy sector. *Energy Policy*, 122, 313–21.
55 Bremmer, I., & Johnston, R. (2009) The rise and fall of resource nationalism. *Survival*, 51(2), 149–58.
56 Klare, M. (2009) *Rising Powers, Shrinking Planet: The New Geopolitics of Energy*. Basingstoke: Macmillan.
57 IRENA, above, n. 50.

58 Bazilian, M., & Ali, S.H. (2018) The peace dividends of energy infrastructure – North Korea, Iran and Beyond. *National Geographic*, August 8.

59 Maresca, J. J. (1995) A 'peace pipeline' to end the Nagorno–Karabakh conflict. *Caspian Crossroads*, 1, 17–18.

60 Shaffer, B. (2013) Natural gas supply stability and foreign policy. *Energy Policy*, 56, 114–25.

61 Fischhendler, I., Herman, L., & Anderman, J. (2016) The geopolitics of cross-border electricity grids: The Israeli–Arab case. *Energy Policy*, 98, 533–43.

62 Colgan, J. D. (2013) Fueling the fire: Pathways from oil to war. *International Security*, 38(2), 147–80.

63 Meierding, E. (2016) Dismantling the oil wars myth. *Security Studies*, 25(2), 258–88.

64 Kelanic, R. A. (2016) The petroleum paradox: Oil, coercive vulnerability, and great power behavior. *Security Studies*, 25(2), 181–213.

65 Klare, M. (2007) *Blood and Oil: The Dangers and Consequences of America's Growing Dependency on Imported Petroleum*. New York: Metropolitan Books.

66 Rovner, J., & Talmadge, C. (2014) Hegemony, force posture, and the provision of public goods: The once and future role of outside powers in securing Persian Gulf oil. *Security Studies*, 23(3), 548–81.

67 Kelanic, above, n. 64.

68 Colgan, J. D. (2010) Oil and revolutionary governments: Fuel for international conflict. *International Organization*, 64(4), 661–94.

69 Colgan, J. D. (2014) Oil, domestic politics, and international conflict. *Energy Research & Social Science*, 1, 198–205.

70 Colgan, J. D. (2013) *Petro-Aggression: When Oil Causes War*. Cambridge: Cambridge University Press.

71 Moran, D., & Russell, J. A. (Eds.) (2008) *Energy Security and Global Politics: The Militarization of Resource Management*. London: Routledge, p. 10.

72 Do, Q. T., Shapiro, J. N., Elvidge, C. D. et al. (2018) Terrorism, geopolitics, and oil security: Using remote sensing to estimate oil production of the Islamic State. *Energy Research & Social Science*, 44, 411–18.

73 Colgan, J. D. (2013) Fueling the fire: Pathways from oil to war. *International Security*, 38(2), 147–80.

74 Gilinsky, V. (2014) Nuclear power, nuclear weapons – Clarifying the links. In Sokolski, H. (Ed.) *Moving beyond Pretense: Nuclear Power and Nonproliferation*. Carlisle, PA: US Army War College Press.

75 Rhodes, R. (1986) *The Making of the Atomic Bomb*. New York: Simon & Schuster.

76 Furhmann, M. (2012) *Atomic Assistance: How "Atoms for Peace" Programs Cause Nuclear Insecurity*. Ithaca, NY: Cornell University Press.

77 Jasanoff, S., & Kim, S.-H. (2009) Containing the atom: Sociotechnical imaginaries and nuclear power in the United States and South Korea. *Minerva*, 47(2), 119–46.

78 Nye, D. E. (2006) *Technology Matters: Questions to Live With.* Cambridge, MA: MIT Press, p. 141.

79 Weart, S. R. (1988) *Nuclear Fear: A History of Images.* Cambridge, MA: Harvard University Press.

80 Ibid., p. 159.

81 Ibid.

82 Busby, R. (1996–1997) The United States' failure to establish a high-level nuclear waste storage facility is threatening its ability to effectively support nuclear nonproliferation. *George Washington Journal of International Law and Economics*, 30, 449–80.

83 Salloum, A. (2008) US–India nuclear deal a non-proliferation disaster. *The Toronto Star*, August 21.

84 Deutch, J. (2007) Priority energy security issues. In Deutch, J., Lauvergeon, A., & Prawiraatmadja, W. (Eds.) *Energy Security and Climate Change.* Washington, DC: Trilateral Commission, pp. 1–50.

85 Bowman, S. (2002) Weapons of mass destruction: The terrorist threat. CRS Report for Congress. Washington, DC: Congressional Research Service, March 7, RL31332.

86 This section was based on IRENA, above, n. 50.

87 Gholz, E., & Press, D. G. (2013) Enduring resilience: How oil markets handle disruptions. *Security Studies*, 22(1), 139–47.

Chapter 4 Energy and the economy: Powering growth and prosperity?

1 Cook, E. (1971) The flow of energy in an industrial society. *Scientific American*, 225(3), 134–47.

2 Kühne, K. (2015) Kings, slaves and the future: Why it is not right to burn fossil fuels and how to stop doing it. Available at http://leave-it-in-the-ground.org/wp-content/uploads/2015/05/Kings-Slaves-and-Future-Script-LINGO.pdf.

3 McCloskey, D. N. (2016) The great enrichment: A humanistic and social scientific account. *Scandinavian Economic History Review*, 64(1), 6–18.

4 Ridley, M. (2010) *The Rational Optimist: How Prosperity Evolves.* New York: Harper.

5 Kühne, above, n. 2.

6 Overland, I. (2016) Energy: The missing link in globalization. *Energy Research & Social Science*, 14, 122–30.

7 Stern, D. I. (2011) The role of energy in economic growth. *Annals of the New York Academy of Sciences*, 1219(1), 26–51. Stern, D. I., & Kander, A. (2012) The role of energy in the industrial revolution and modern economic growth. *The Energy Journal*, 33(3), 125–52.

8 Rühl, C. (2010) Global energy after the crisis: Prospects and priorities. *Foreign Affairs*, 89(2), 63–75.

9 Bressand, A. (2013) The role of markets and investment in global energy. In Goldthau, A. (Ed.) *The Handbook of Global Energy Policy*. Chichester: John Wiley & Sons, pp. 15–29.

10 England, J., Bean, G., & Mittal, A. (2015) Following the capital trail in oil and gas. A report by the Deloitte Center for Energy Solution. London: Deloitte University Press.

11 CNN Money (2012) 10 most expensive energy projects in the world. *CNN Money*, August 27. Available at https://money.cnn.com/gallery/news/economy/2012/08/27/expensive-energy-projects/10.html.

12 IEA (2019) *World Energy Investment: 2019 edition*. Paris: IEA/OECD.

13 Höök, M., Söderbergh, B., Jakobsson, K., & Aleklett, K. (2009) The evolution of giant oil field production behavior. *Natural Resources Research*, 18(1), 39–56.

14 Bressand, above, n. 9.

15 IEA (2015) *World Energy Outlook 2015*. Paris: OECD/IEA, p. 117.

16 Mitchell, J. V., & Mitchell, B. (2015) States and markets in the oil industry. In Belyi, A. V., & Talus, K. (Eds.) *States and Markets in Hydrocarbon Sectors*. London: Palgrave Macmillan, pp. 17–39.

17 Högselius, P. (2019) *Energy and Geopolitics*. New York: Routledge, pp. 69–70.

18 UNCTAD (2014) Key trends in international merchandise trade. Geneva: UNCTAD, p. 10.

19 Högselius, above, n. 17.

20 Iran inches closer to dream of gasoline independence. *Reuters*, March 21, 2018.

21 Koh, A., Wittels, J., & Sundria, S. (2019) Venezuela needs to dilute its oil. That's going to become harder. *Bloomberg*, January 29.

22 Bressand, above, n. 9.

23 Baev, P. K., & Øverland, I. (2010) The South Stream versus Nabucco pipeline race: Geopolitical and economic (ir)rationales and political stakes in mega-projects. *International Affairs*, 86(5), 1075–90.

24 Bressand, above, n. 9, p. 17.

25 Barma, N., Kaiser, K., Le, T. M., & Viñuela, L. (2011) *Rents to Riches? The Political Economy of Resource-led Development*. Washington, DC: The World Bank, p. ix.

26 Luciani, G. (1990), Allocation vs production states: A theoretical framework. In Luciani, G. (Ed.), *The Arab State*. London: Routledge, pp. 65–84.

27 Ruta, M., & Venables, A. J. (2012) International trade in natural resources: Practice and policy. *Annual Review of Resource Economics*, 4(1), 331–52.

28 OPEC (2018) Who gets what from imported oil? Available at https://www.opec.org/opec_web/en/publications/341.htm.

29 European Commission (2019) Energy prices and costs in Europe. COM(2019) 1 final. Brussels: European Commission.

30 IEA (2018) *Key World Energy Statistics*. Paris: IEA/OECD.

31 Rühl, C. (2010) Global energy after the crisis: Prospects and priorities. *Foreign Affairs*, 89(2), 63–75.

32 Hirst, N. (2018) *The Energy Conundrum: Climate Change, Global Prosperity and the Tough Choices We Have to Make*. Singapore: World Scientific Publishing, pp. 67 and 133.

33 Davis, S. J., & Caldeira, K. (2010) Consumption-based accounting of CO2 emissions. *Proceedings of the National Academy of Sciences*, 107(12), 5687–92.

34 IEA (2018) *World Energy Outlook 2018*. Paris: OECD/IEA.

35 National Bureau of Statistics of China (2019) *China Statistical Yearbook 2019*. Available at http://www.stats.gov.cn/tjsj/ndsj/2018/indexeh.htm.

36 Green, F., & Stern, N. (2017) China's changing economy: Implications for its carbon dioxide emissions. *Climate Policy*, 17(4), 423–42.

37 Aguilera, R. F., & Radetzki, M. (2015) *The Price of Oil*. Cambridge: Cambridge University Press, pp. 1–2.

38 Hamilton, J. D. (2008) Oil and the macroeconomy. *The New Palgrave Dictionary of Economics*. Basingstoke: Palgrave Macmillan.

39 WTO (2019) Merchandise imports by product group – annual (million US dollar). Available at https://data.wto.org/.

40 IMF (2000) The Impact of Higher Oil Prices on the Global Economy. Washington, DC: IMF.

41 European Commission (2016) *Energy Prices and Costs in Europe*. COM(2016) 769 final. Brussels: European Commission.

42 Cunningham, S. (2019) Anti-OPEC Bill allowing US to sue oil cartel moves forward. *Bloomberg*, February 7.

43 Colgan, J. D. (2014) The emperor has no clothes: The limits of OPEC in the global oil market. *International Organization*, 68(3), 599–632.

44 Dale, S. (2015) New economics of oil. Paper presented to Society of Business Economists Annual Conference, London, October 13. Available at https://www.bp.com/content/dam/bp/business-sites/en/global/corporate/pdfs/news-and-insights/speeches/new-economics-of-oil-spencer-dale.pdf.

45 G20 Leaders' Statement (2009) Pittsburgh Summit, September 24–25. Available at http://www.g20.utoronto.ca/2009/2009communique0925.html.

46 Friends of Fossil Fuel Subsidy Reform (n.d.) The top 5 reasons for reform. Available at http://fffsr.org/reasons-for-reform/.

47 IEA (2018) *World Energy Outlook*. Paris: OECD/IEA, pp. 110–11.

48 Coady, D., Parry, I., Sears, L., & Shang, B. (2017) How large are global fossil fuel subsidies? *World Development*, 91, 11–27.

49 Sovacool, B. K. (2017) Reviewing, reforming, and rethinking global energy subsidies: Towards a political economy research agenda. *Ecological Economics*, 135, 150–63.

50 OECD (2018) *OECD Companion to the Inventory of Support Measures for Fossil Fuels 2018*. Paris: OECD Publishing, p. 15.

51 Rentschler, J., & Bazilian, M. (2017) Reforming fossil fuel subsidies: Drivers, barriers and the state of progress. *Climate Policy*, 17(7), 891–914.

52 Krane, J. (2019) *Energy Kingdoms: Oil and Political Survival in the Persian Gulf*. New York: Columbia University Press.

53 Inchauste, G., & Victor, D. G. (2017) *The Political Economy of Energy Subsidy Reform*. Washington, DC: World Bank.

54 Lockwood, M. (2015) Fossil fuel subsidy reform, rent management and political fragmentation in developing countries. *New Political Economy*, 20(4), 475–94.

55 Ross, M. L., Hazlett, C., & Mahdavi, P. (2017) Global progress and backsliding on gasoline taxes and subsidies. *Nature Energy*, 2(1), 16201.

56 IEA, above, n. 47, p. 255.

57 IEA (2014) *World Energy Outlook 2014*. Paris: IEA.

58 Fortune, http://fortune.com/global500/list/.

59 World Bank (2019) GPD (current US$). Available at https://data.world bank.org/indicator/ny.gdp.mktp.cd?most_recent_value_desc=true.

60 Coll, S. (2012) *Private Empire: ExxonMobil and American Power*. Harmondsworth: Penguin.

61 Downie, C. (2019) *Business Battles in the US Energy Sector*. New York: Routledge.

62 Colgan, J. D., & Van de Graaf, T. (2017) A crude reversal: The political economy of the United States crude oil export policy. *Energy Research & Social Science*, 24, 30–35.

63 Tran, M. (2004) Shell fined over reserves scandal. *Guardian*, July 29.

64 Gross, A., & Walker, O. (2019) Shell faces pressure on gay rights over Brunei venture. *Financial Times*, April 25.

65 Jordan, G. (2001) *Shell, Greenpeace and the Brent Spar*. Basingstoke: Palgrave Macmillan.

66 Vaughan, A. (2018) BP's Deepwater Horizon bill tops $65bn. *Guardian*, January 16.

67 Setzer, J., & Vanhala, L. (2019) Climate change litigation: A review of research on courts and litigants in climate government. *WIREs Climate Change*, 10, e580.

68 Oreskes, N., & Conway, E. M. (2011) *Merchants of Doubt: How a Handful of Scientists Obscured the Truth on Issues from Tobacco Smoke to Global Warming*. New York: Bloomsbury Publishing.

69 Olszynski, M., Mascher, S., & Doelle, M. (2017) From smokes to smokestacks: Lessons from tobacco for the future of climate change liability. *Georgetown Environmental Law Review*, 30, 1–45.

70 Joint statement between institutional investors on behalf of Climate Action 100+ and Royal Dutch Shell plc (Shell), December 3, 2018. Available at https://www.shell.com/media/news-and-media-releases/2018/joint-statement-between-institutional-investors-on-behalf-of-climate-action-and-shell.html.

71 Hume, N., Sheppard, D., & Sanderson, H. (2019) Glencore vows to cap global coal production. *Financial Times*, February 20.

72 Nasiritousi, N. (2017) Fossil fuel emitters and climate change: Unpacking the governance activities of large oil and gas companies. *Environmental Politics*, 26(4), 621–47. Bach, M. (2019) The oil and gas sector: From climate laggard to climate leader? *Environmental Politics*, 28(1), 87–103.

73 Crooks, E., & Raval, A. (2019) Shell aims to become world's largest electricity company. *Financial Times*, March 13.

74 Skjærseth, J. B., & Skodvin, T. (2001) Climate change and the oil industry: Common problems, different strategies. *Global Environmental Politics*, 1(4), 43–64.

75 Crooks, E., & Stacey, K. (2016) Big Oil: From Black to Green. *Financial Times*, June 28.

76 Helm, D. (2017) *Burn Out: The Endgame for Fossil Fuels*. New Haven, CT: Yale University Press, p. 204.

77 IRENA (2018) *Corporate Sourcing of Renewable Energy: Market and Industry Trends*. Abu Dhabi: International Renewable Energy Agency.

78 RE100 website: http://there100.org/.

79 Tordo, S. (2011) *National Oil Companies and Value Creation*. Washington, DC: The World Bank.

80 Reed, S. (2019) Saudi Aramco Is World's Most Profitable Company, Beating Apple by Far. *New York Times*, April 1.

81 Parraga, M. (2019) Exclusive: After US sanctions, Venezuela seeks to collect some oil payments via Rosneft. *Reuters*, April 18.

82 Downs, E. S. (2007) The fact and fiction of Sino-African energy relations. *China Security*, 3(3), 42–86. IEA (2011) *Overseas Investments by Chinese National Oil Companies*. Paris: OECD/IEA.

83 Victor, D. G., Hults, D. R., & Thurber, M. C. (Eds.) (2011) *Oil and Governance: State-Owned Enterprises and the World Energy Supply*. Cambridge: Cambridge University Press. Marcel, V. (2006) *Oil Titans: National Oil Companies in the Middle East*. Washington, DC: Brookings Institution Press.

84 Auty, R. (1993) *Sustaining Development in Mineral Economies: The Resource Curse Thesis*. New York: Routledge.

85 Gochberg, W., & Menaldo, V. (2016) The resource curse puzzle across four waves of work. In Van de Graaf, T. et al. (Eds.) *The Palgrave Handbook of the International Political Economy of Energy*. London: Palgrave Macmillan, pp. 505–25.

86 Sachs, J. D., & Warner, A. M. (2001) The curse of natural resources. *European Economic Review*, 45(4–6), 827–38.

87 Bulte, E. H., Damania, R., & Deacon, R. T. (2005) Resource intensity, institutions, and development. *World Development*, 33(7), 1029–44.

88 Gylfason, T. (2001) Natural resources, education, and economic development. *European Economic Review*, 45, 847–59.

89 Ross, M. L. (2012) *The Oil Curse: How Petroleum Wealth Shapes the Development of Nations.* Princeton, NJ: Princeton University Press.

90 Arezki, R., & Brückner, M. (2011) Oil rents, corruption, and state stability: Evidence from panel data regressions. *European Economic Review*, 55(7), 955–63.

91 Ross, M. L. (2001) Does oil hinder democracy? *World Politics*, 53(3), 325–61.

92 Collier, P., & Hoeffler, A. (2004) Greed and grievance in civil war. *Oxford Economic Papers*, 56(4), 563–95.

93 Rosser, A. (2006) The political economy of the resource curse: A literature survey. Working paper series, 268. Brighton: IDS.

94 Ross, M. (2006) A closer look at oil, diamonds, and civil war. *Annual Review of Political Science*, 9, 265–300. Ross, M. L. (2015) What have we learned about the resource curse? *Annual Review of Political Science*, 18, 239–59.

95 Menaldo, V. (2016) *The Institutions Curse: Natural Resources, Politics, and Development.* New York: Cambridge University Press.

96 Gilberthorpe, E., & Papyrakis, E. (2015) The extractive industries and development: The resource curse at the micro, meso and macro levels. *The Extractive Industries and Society*, 2(2), 381–90.

97 Frankel, J. A. (2012) The natural resource curse: A survey of diagnoses and some prescriptions. HKS Faculty Research Working Paper Series RWP12-014, John F. Kennedy School of Government, Harvard University.

98 Ross, above, n. 89.

99 Tsalik, S. (2003) The hazards of petroleum wealth. In Ebel, R. (Ed.) *Caspian Oil Windfalls: Who Will Benefit?* Washington, DC: Open Society Institute, pp. 1–15.

100 Robinson, J. A., & Torvik, R. (2005) White elephants. *Journal of Public Economics*, 89(2–3), 197–210.

101 Ross, above, n. 89, pp. 44–5.

102 Ibid., p. 45.

103 Ross, What have we learned about the resource curse?, above, n. 94.

104 Mahdavi, H. (1970) The patterns and problems of economic development in rentier states. In Cook, M. A. (Ed.) *Studies in the Economic History of the Middle East.* London: Oxford University Press. Beblawi, H., & Luciani, G. (Eds.) (1987) *The Rentier State.* New York: Routledge.

105 Ross, above, n. 91.

106 Ross, What have we learned about the resource curse?, above, n. 94.
107 Mehlum, H., Moene, K., & Torvik, R. (2006) Institutions and the resource curse. *The Economic Journal*, 116(508), 1–20.
108 Bulte et al., above, n. 87.
109 Gilberthorpe & Papyrakis, above, n. 96.
110 Robinson, J. A., Torvik, R., & Verdier, T. (2006) Political foundations of the resource curse. *Journal of Development Economics*, 79(2), 447–68.
111 El-Katiri, L. (2014) The guardian state and its economic development model. *Journal of Development Studies*, 50(1), 22–34.
112 Ross, above, n. 89.
113 Collier, P., & Hoeffler, A. (2004) Greed and grievance in civil war. *Oxford Economic Papers*, 56(4), 563–95.
114 Ross, What have we learned about the resource curse?, above, n. 94.
115 Ross, above, n. 89.
116 Ibid.
117 Collier, P. (2010) *The Plundered Planet: Why We Must – And How We Can – Manage Nature for Global Prosperity*. Oxford: Oxford University Press.
118 Humphreys, M., & Sandbu, M. E. (2007) The political economy of natural resource funds. In Humphreys, M., Sachs, J. D., & Stiglitz, J. E. (Eds.) *Escaping the Resource Curse*. New York: Columbia University Press, pp. 194–233.
119 Frankel, above, n. 97.
120 Humphreys, M., Sachs, J. D., & Stiglitz, J. E. (2007) Future directions for the management of natural resources. In Humphreys, M., Sachs, J. D., & Stiglitz, J. E. (Eds.) *Escaping the Resource Curse*. New York: Columbia University Press, pp. 322–36.
121 Van Alstine, J., & Andrews, N. (2016) Corporations, civil society, and disclosure: A case study of the extractive industries transparency initiative. In Van de Graaf, T. et al. (Eds) *The Palgrave Handbook of the International Political Economy of Energy*. London: Palgrave Macmillan, pp. 95–114.
122 Van de Graaf, T., & Bradshaw, M. (2018) Stranded wealth: Rethinking the politics of oil in an age of abundance. *International Affairs*, 94(6), 1309–28.
123 Hotelling, H. (1931) The economics of exhaustible resources. *Journal of Political Economy*, 39(2), 137–75.
124 Van de Graaf, T., & Verbruggen, A. (2015) The oil endgame: Strategies of oil exporters in a carbon-constrained world. *Environmental Science & Policy*, 54, 456–62.
125 Van der Ploeg, F. (2016) Fossil fuel producers under threat. *Oxford Review of Economic Policy*, 32(2), 206–22.
126 Cairns, R. D. (2014) The green paradox of the economics of exhaustible resources. *Energy Policy*, 65, 78–85.
127 Van der Ploeg, above, n. 125.

128 Hartwick, J. M. (1977) Intergenerational equity and the investing of rents from exhaustible resources. *American Economic Review*, 67(5), 972–74.

129 World Bank (2011) *The Changing Wealth of Nations: Measuring Sustainable Development in the New Millennium*. Washington, DC: World Bank, p. 11.

130 Baffes, J., Kose, M. A., Ohnsorge, F., & Stocker, M. (2015) The great plunge in oil prices: Causes, consequences, and policy responses. World Bank Group Policy Research Note.

131 Van de Graaf, T. (2018) Battling for a shrinking market: Oil producers, the renewables revolution and the risk of stranded assets. In Scholten, D. (Ed.) *The Geopolitics of Renewables*. New York: Springer, pp. 97–121.

132 Helm, above, n. 76.

133 Ross, M. L. (2017) What do we know about economic diversification in oil-producing countries? Available at SSRN: http://dx.doi.org/10.2139/ssrn.3048585.

134 Ross, above, n. 89.

135 Gelb, A. (2010) Economic diversification in resource rich countries. Available at https://www.imf.org/external/np/seminars/eng/2010/afrfin/pdf/Gelb2.pdf.

136 Tordo, above, n. 79.

Chapter 5   Energy and the environment: Wrecking the planet?

1 A nice flowchart of the sectors and activities in the world economy contributing to greenhouse gas emissions can be found at the website of the World Resources Institute: http://pdf.wri.org/world_greenhouse_gas_emissions_flowchart.pdf.

2 Nugent, C. (2019) Carbon dioxide concentration in the Earth's atmosphere has hit levels unseen for 3 million years. *Time*, May 14.

3 IPCC (2018) Global warming of 1.5°C. An IPCC Special Report on the impacts of global warming of 1.5°C above pre-industrial levels and related global greenhouse gas emission pathways, in the context of strengthening the global response to the threat of climate change, sustainable development, and efforts to eradicate poverty. Geneva: IPCC.

4 UNEP (2018) Emissions Gap Report. Nairobi: UNEP.

5 Winkelmann, R., Levermann, A., Ridgwell, A., & Caldeira, K. (2015) Combustion of available fossil fuel resources sufficient to eliminate the Antarctic Ice Sheet. *Science Advances*, 1(8), e1500589.

6 Steffen, W., Rockström, J., Richardson, K., et al. (2018) Trajectories of the Earth system in the Anthropocene. *Proceedings of the National Academy of Sciences*, 115(33), 8252–9.

7 World Health Organization (2018) How air pollution is destroying our health. Available at https://www.who.int/air-pollution/news-and-events/how-air-pollution-is-destroying-our-health.

8 International Energy Agency (2016) *World Energy Outlook 2016*. Paris: OECD/IEA.

9 Ibid.

10 US EPA (2006) *Particulate Matter: Health and Environment*.

11 World Health Organization (2006) *Fuel for Life*. Geneva: WHO, p. 8.

12 Ibid.

13 Masud, J., Sharan, D., & Lohani, B. N. (2007) *Energy for All: Addressing the Energy, Environment, and Poverty Nexus in Asia*. Manila: Asian Development Bank.

14 Sovacool, B. K., & Sovacool, K. E. (2009) Identifying future electricity water tradeoffs in the United States. *Energy Policy*, 37(7), 2763–73.

15 International Energy Agency (2018) *World Energy Outlook 2018*. Paris: OECD/IEA.

16 Dale, V. H., Efroymson, R. A., & Kline, K. L. (2011) The land use–climate change–energy nexus. *Landscape Ecology*, 26, 755–73.

17 Girardet, H., & Mendonca, M. (2009) *A Renewable World: Energy, Ecology, Equality*. Totnes, Devon: Green Books, p. 188.

18 Streck, C. (2008) Forests, carbon markets, and avoided deforestation: Legal implications. *Carbon & Climate Law Review*, 2(3), 239–47.

19 Streck, C., Pedroni, L., Porrua, M., & Dutschke, M. (2008) Creating incentives for avoiding further deforestation: The nested approach. In Streck, C., O'Sullivan, R., Janson-Smith, T., & Tarasofsky, R. G. (Eds.) *Climate Change and Forests: Emerging Policy and Market Opportunities*. Washington, DC: Brookings Institution Press, pp. 237–49.

20 United Nations Food and Agricultural Organization (2006) *Global Forest Resource Assessment*. Rome: FAO.

21 Boucher, D. (2008) *Out of the Woods: A Realistic Role for Tropical Forests in Curbing Global Warming*. Washington, DC: Union of Concerned Scientists.

22 Holdren, J. P., & Smith, K. R. (2000) Energy, the environment, and health. In Kjellstrom, T., Streets, D., & Wang, X. (Eds.) *World Energy Assessment: Energy and the Challenge of Sustainability*. New York: United Nations Development Programme, pp. 61–110.

23 Clayton, B. C. (2015) *Market Madness: A Century of Oil Panics, Crises, and Crashes*. New York: Oxford University Press.

24 Jevons, W. S. (1866) *The Coal Question: An Inquiry Concerning the Progress of the Nation, and the Probable Exhaustion of Our Coal-Mines*. London: Macmillan, p. 374.

25 Miller, R. G., & Sorrell, S. R. (2014) The future of oil supply. *Philosophical Transactions of the Royal Society A: Mathematical, Physical and Engineering Sciences*, 372(2006).

26 Aleklett, K. (2006) Oil: A bumpy road ahead. *World Watch Magazine*, January/February, pp. 10–12.

27 Hicks, B., & Nelder, C. (2008) *Profit from the Peak: The End of Oil and the Greatest Investment Event of the Century*. New York: John Wiley & Sons.

28 International Energy Agency (2008) *World Energy Outlook 2008*. Paris: OECD/IEA, p. 3.
29 US Energy Information Administration (EIA) (2018) US field production of crude oil. Available at https://www.eia.gov/dnav/pet/hist/LeafHandler.ashx?n=PET&s=MCRFPUS2&f=A.
30 Dale, S. (2015) New economics of oil. Paper presented to Society of Business Economists Annual Conference, London, October 13. Available at https://www.bp.com/content/dam/bp/business-sites/en/global/corporate/pdfs/news-and-insights/speeches/new-economics-of-oil-spencer-dale.pdf.
31 International Energy Agency (2017) *World Energy Outlook 2017*. Paris: OECD/IEA, p. 26.
32 Harvey, H., Orr, Jr., F. M., & Vondrich, C. (2013) A trillion tons. *Daedalus*, 142(1), 8–25.
33 McKibben, B. (2012) Global warming's terrifying new math. *Rolling Stone*, July 19, pp. 32–44.
34 Leaton, J. (2011) Unburnable carbon – Are the world's financial markets carrying a carbon bubble. *Carbon Tracker Initiative*, 105, 16.
35 McGlade, C., & Ekins, P. (2015) The geographical distribution of fossil fuels unused when limiting global warming to 2°C. *Nature*, 517(7533), 187–90.
36 Millar, R., Allen, M., Rogelj, J., & Friedlingstein, P. (2016) The cumulative carbon budget and its implications. *Oxford Review of Economic Policy*, 32(2), 323–42.
37 Ruhl, C. (2019) The war on plastic will dent oil demand more than anticipated. *Financial Times*, February 17.
38 Ellen MacArthur Foundation (2016) *The New Plastics Economy: Rethinking the Future of Plastics*. Available at https://www.ellenmacarthurfoundation.org/publications/the-new-plastics-economy-rethinking-the-future-of-plastics. Jefferson, M. (2019) Whither plastics? – Petrochemicals, plastics and sustainability in a garbage-riddled world. *Energy Research & Social Science*, 56, 101229.
39 Van de Graaf, T. & Blondeel, M. (2018) Fossil fuel subsidy reform: An international norm perspective. In Skovgaard, J., & van Asselt, H. (Eds.) *The Politics of Fossil Fuel Subsidies and their Reform*. Cambridge: Cambridge University Press.
40 Meckling, J., & Nahm, J. (2019) The politics of technology bans: Industrial policy competition and green goals for the auto industry. *Energy Policy*, 126, 470–9.
41 Gupta, J. (2014) *The History of Global Climate Governance*. Cambridge: Cambridge University Press.
42 Damro, C., & Méndez, P. L. (2003) Emissions trading at Kyoto: From EU resistance to Union innovation. *Environmental Politics*, 12(2), 71–94.
43 Anderson, K., & Bows, A. (2011) Beyond "dangerous" climate change: Emission scenarios for a new world. *Philosophical Transactions*

of the Royal Society of London A: Mathematical, Physical and Engineering Sciences, 369(1934), 20–44.

44 Gupta, above, n. 41.

45 Shishlov, I., Morel, R., & Bellassen, V. (2016) Compliance of the parties to the Kyoto Protocol in the first commitment period. *Climate Policy*, 16(6), 768–82.

46 Betz, R., & Sato, M. (2006) Emissions trading: Lessons learnt from the 1st phase of the EU ETS and prospects for the 2nd phase. *Climate Policy*, 6(4), 351–9.

47 Schneider, L. (2009) Assessing the additionality of CDM projects: Practical experiences and lessons learned. *Climate Policy*, 9(3), 242–54.

48 UNFCCC (2015) Paris Agreement as contained in the report of the Conference of the Parties on its twenty-first session, FCCC/CP/2015/10/Add.1.

49 Falkner, R. (2016) The Paris Agreement and the new logic of international climate politics. *International Affairs*, 92(5), 1107–25.

50 Urpelainen, J., & Van de Graaf, T. (2018) United States non-cooperation and the Paris Agreement. *Climate Policy*, 18(7), 839–51.

51 Corner, A., & Pidgeon, N. (2014) Geoengineering, climate change scepticism and the 'moral hazard' argument: An experimental study of UK public perceptions. *Philosophical Transactions of the Royal Society A: Mathematical, Physical and Engineering Sciences*, 372(2031), 20140063.

52 Hale, T. (2016) "All hands on deck": The Paris Agreement and non-state climate action. *Global Environmental Politics*, 16(3), 12–22.

53 Bulkeley, H., Andonova, L. B., Betsill, M. M., et al. (2014) *Transnational Climate Change Governance*. Cambridge: Cambridge University Press.

54 www.c40.org.

55 Bernstein, S., & Hoffmann, M. (2018) The politics of decarbonization and the catalytic impact of subnational climate experiments. *Policy Sciences*, 51(2), 189–211.

56 http://there100.org/companies.

57 Green, F. (2018) Anti-fossil fuel norms. *Climatic Change*, 150(1–2), 103–16.

58 Green, F., & Denniss, R. (2018) Cutting with both arms of the scissors: The economic and political case for restrictive supply-side climate policies. *Climatic Change*, 150(1–2), 73–87.

59 Newell, P., & Simms, A. (2019) Towards a fossil fuel non-proliferation treaty. *Climate Policy*, https://doi.org/10.1080/14693062.2019.1636759.

Chapter 6 Energy and justice: Equitable and fair?

1 Matthew, R. A. (2007) Climate change and human security. In DiMento, J. F. C., & Doughman, P. (Eds.) *Climate Change: What It Means for Us, Our Children, and Our Grandchildren*. Cambridge, MA: MIT Press, pp. 161–80.

2 Wilkinson, P., Smith, K., Joffe, M., & Haines, A. (2007) A global perspective on energy: Health effects and injustices, *Lancet*, 370, 965–77.
3 Jenkins, K., McCauley, D., Heffron, R., Stephan, H., & Rehner, R. (2016) Energy justice: A conceptual review. *Energy Research & Social Science*, 11, 174–82. Sovacool, B. K., & Dworkin, M. H. (2014) *Global Energy Justice: Problems, Principles, and Practices*. Cambridge: Cambridge University Press. Sovacool, B. K., Burke, M., Baker, L., Kotikalapudi, C. K., & Wlokas, H. (2017) New frontiers and conceptual frameworks for energy justice. *Energy Policy*, 105, 677–91.
4 Dominish, E., Briggs, C., Teske, S., & Mey, F. (2019) Just transition: Employment projections for the 2.0°C and 1.5°C scenarios. In Teske, S. (Ed.) *Achieving the Paris Climate Agreement Goals*. London: Springer, pp. 413–35. Healy, N., & Barry, J. (2017) Politicizing energy justice and energy system transitions: Fossil fuel divestment and a "just transition." *Energy Policy*, 108, 451–9. Heffron, R. J., & McCauley, D. (2018) What is the "just transition"? *Geoforum*, 88, 74–7.
5 World Bank (2018) Boosting access to electricity in Africa through innovation, better regulation. *Africa's Pulse*, 17.
6 World Bank (2017) *Getting Electricity: Factors Affecting the Reliability of Electricity Supply*. Doing Business 2017. Washington, DC: World Bank.
7 International Energy Agency (2017) *World Energy Outlook 2017 – Special Report: Energy Access Outlook*.
8 Bazilian, M. D. (2015) Power to the poor. *Foreign Affairs*, 94(2).
9 Patrick, E. (2007) Sexual violence and firewood collection in Darfur. *Forced Migration Review*, 27, 40–1.
10 International Energy Agency (2017) *World Energy Outlook 2017 – Special Report: Energy Access Outlook, p. 84.
11 O'Brien, M. (2011) Fuel poverty in England. *The Lancet*, December 5.
12 Bouzarovski, S., & Petrova, S. (2015) A global perspective on domestic energy deprivation: Overcoming the energy poverty–fuel poverty binary. *Energy Research & Social Science*, 10, 31–40.
13 See EU Fuel Poverty Network at http://www.fuelpoverty.eu.
14 Public Health England (2014) *Local Action on Health Inequalities: Fuel Poverty and Cold Home-Related Health Problems*. London: Public Health England. National Audit Office (2003) Warm front: Helping to combat fuel poverty. Report by the Comptroller and Auditor General, HC 769 Session 2002–2003: 25 June, p. 1. See also O'Brien, above, n. 11; Rudge, J., & Gilchrist, R. (2005) Excess winter morbidity among older people at risk of cold homes: A population-based study in a London borough. *Journal of Public Health*, 27(4), 353–8.
15 National Audit Office, above, n. 14, p. 1.
16 Hutton, S., & Hardman, G. (1993) *Assessing the Impact of VAT on Fuel on Low-Income Households: Analysis of the Fuel Expenditure Data from the 1991 Family Expenditure Survey*. York: Social Policy Research Unit, University of York.

17 Walker, G. (2008) Decentralised systems and fuel poverty: Are there any links or risks? *Energy Policy*, 36, 4514–17.

18 Guertler, P., & Smith, P. (2018) Cold homes and excess winter deaths: A preventable public health epidemic that can no longer be tolerated. National Energy Action and E3G. Available at https://www. nea.org.uk/wp-content/uploads/2018/02/E3G-NEA-Cold-homes-and-excess-winter-deaths.pdf.

19 Liddell, C., Morris, C., Thomson, H., & Guiney, C. (2016) Excess winter deaths in 30 European countries 1980–2013: A critical review of methods. *Journal of Public Health*, 38(4), 806–14.

20 van Veelen, B., & van der Horst, D. (2018) What is energy democracy? Connecting social science energy research and political theory. *Energy Research & Social Science*, 46, 19–28.

21 Szulecki, K. (2018) Conceptualizing energy democracy. *Environmental Politics*, 27(1), 21–41. Burke, M. J., & Stephens, J. C. (2017) Energy democracy: Goals and policy instruments for sociotechnical transitions. *Energy Research & Social Science*, 33, 35–48.

22 Ballard, C., & Banks, G. (2003) Resource wars: The anthropology of mining. *Annual Review of Anthropology*, 32, 287–313, at p. 305.

23 Watts, M. J. (2005) Righteous oil: Human rights, the oil complex, and corporate social responsibility. *Annual Review of Environment and Resources*, 30, 373–407.

24 Maximus, C. (2011) Islamic Republic corruption scandal: $11 billion in oil money missing. March 28. http://iranchannel.org/archives/962 (no longer available).

25 Downie, C. (2017) Business actors, political resistance, and strategies for policymakers. *Energy Policy*, 108, 583–92.

26 InfluenceMap (2019) *Big Oil's Real Agenda on Climate Change*. London: InfluenceMap.

27 Hansen, J. (2008) Global warming twenty years later: Tipping points near. Paper presented at the National Press Club, Washington, DC, June 23. Supran, G., & Oreskes, N. (2017) Assessing ExxonMobil's climate change communications (1977–2014). *Environmental Research Letters*, 12, 084019.

28 Brulle, R. J. (2014) Institutionalizing delay: Foundation funding and the creation of US climate change counter-movement organizations. *Climatic Change*, 122(4), 681–94.

29 Brulle, R. J., Turner, L. H., Carmichael, J., & Jenkins, J. C. (2007) Measuring social movement organization populations: A comprehensive census of US environmental movement organizations. *Mobilization: An International Quarterly Review*, 12(3), 195–211. Farrell, J. (2016) Corporate funding and ideological polarization about climate change. *Proceedings of the National Academy of Sciences*, 113(1), 92–7. Farrell, J. (2016) Network structure and influence of the climate change counter-movement. *Nature Climate Change*, 6(4), 370–4. Jacques, P. J.,

Dunlap, R. E., & Freeman, M. (2008) The organisation of denial: Conservative think tanks and environmental scepticism. *Environmental Politics*, 17(3), 349–85. McCright, A. M., & Dunlap, R. E. (2003) Defeating Kyoto: The Conservative movement's impact on US climate change policy. *Social Problems*, 50(3), 348–73.

30 Eisgruber, L. (2013) The resource curse: Analysis of the applicability to the large-scale export of electricity from renewable sources. *Energy Policy*, 57, 429–40.

31 Walker, C., & Baxter, J. (2017) Procedural justice in Canadian wind energy development: A comparison of community-based and technocratic siting processes. *Energy Research & Social Science*, 29, 160–9.

32 Wheatley, S., Sovacool, B. K., & Sornette, D. (2017) Of disasters and dragon kings: A statistical analysis of nuclear power incidents & accidents. *Risk Analysis*, 37(1), 99–115.

33 Del Sesto, S. L. (1979) *Science, Politics, and Controversy: Civilian Nuclear Power in the United States, 1946–1974*. Boulder, CO: Westview Press.

34 Rose, D. J. (1981) Energy and history. *American Heritage*, 32, 79–80.

35 Ramana, M. V., & Rao, D. B. (2010) The environmental impact assessment process for nuclear facilities: An examination of the Indian experience. *Environmental Impact Assessment Review*, 30, 268–71.

36 Ramana, M. V. (2009) Nuclear power: Economic, safety, health, and environmental issues of near-term technologies. *Annual Review of Environment and Resources*, 34, 127–52.

37 de Saillan, C. (2010) Disposal of spent nuclear fuel in the United States and Europe: A persistent environmental problem. *Harvard Environmental Law Review*, 34, 462–519.

38 Sovacool, B. K. (2011) *Contesting the Future of Nuclear Power: A Critical Global Assessment of Atomic Energy*. London: World Scientific.

39 Marples, D. R., & Ludwig, G. (2008) Fallout: The enduring tragedy of Chernobyl. In Cohen, D. E. (Ed.) *What Matters: The World's Preeminent Photojournalists and Thinkers Depict Essential Issues of our Time*. New York: Sterling Press, pp. 67–81.

40 Walsh, S., & Stainsby, M. (2010) The smell of money: Alberta's tar sands. In Abramsky, K. (Ed.) *Sparking a Worldwide Energy Revolution: Social Struggles in the Transition to a Post-Petrol World*. Oakland, CA: AK Press, pp. 333–44.

41 Cheon, A., & Urpelainen, J. (2018) *Activism and the Fossil Fuel Industry*. Abingdon: Routledge.

42 Levin, S. (2017) Dakota Access pipeline activists say police have used "excessive" force. *The Guardian*, January 18. Available at https://www.theguardian.com/us-news/2017/jan/18/dakota-access-pipeline-protesters-police-used-excessive-force.

43 Oceransky, S., (2010) Fighting the enclosure of wind: Indigenous resistance to the privatization of the wind resource in Southern Mexico. In Abramsky, K. (Ed.) *Sparking a Worldwide Energy Revolution: Social*

Struggles in the Transition to a Post-Petrol World. Oakland, CA: AK Press, pp. 505–22.

44 Sovacool, B. K., Sidortsov, R., & Jones, B. (2014) *Energy Security, Equality, and Justice*. London: Routledge.

45 Ibid.

46 Watts, above, n. 23.

47 Pilkington, E. (2009) Shell pays out $15.5m over Saro-Wiwa killing. *The Guardian*, June 9. Available at https://www.theguardian.com/world/2009/jun/08/nigeria-usa.

48 Sovacool, B. K. (2019) Toxic transitions in the lifecycle externalities of a digital society: The complex afterlives of electronic waste in Ghana. *Resources Policy*, 64, 101459.

49 Sovacool, B. K. (2019) The precarious political economy of cobalt: Balancing prosperity, poverty, and brutality in artisanal and industrial mining in the Democratic Republic of the Congo. *Extractive Industries & Society*, 6(3), 915–39.

50 World Commission on Dams (2000) *Dams and Development: A New Framework for Decision-making. The Report of the World Commission on Dams*. London: Earthscan.

51 Sovacool, B. K., & Bulan, L. C. (2011) Behind an ambitious megaproject in Asia: The history and implications of the Bakun hydroelectric dam in Borneo. *Energy Policy*, 39(9), 4842–59.

52 Sovacool, B. K., & Bulan, L. C. (2012) Energy security and hydropower development in Malaysia: The drivers and challenges facing the Sarawak corridor of renewable energy (SCORE). *Renewable Energy*, 40(1), 113–29; Sovacool, B. K., & Bulan, L. C. (2011) Meeting targets, missing people: The energy security implications of the Sarawak corridor of renewable energy (SCORE) in Malaysia. *Contemporary Southeast Asia*, 33(1), 56–82; Sovacool, B. K., & Bulan, L. C. (2013) They'll be dammed: The sustainability implications of the Sarawak corridor of renewable energy (SCORE) in Malaysia. *Sustainability Science*, 8(1), 121–33.

53 Myers, S. L. (2010) Lament for a once-lovely waterway. *New York Times*, June 12; Myers, S. L. (2010) Vital river is withering, and Iraq has no answer. *New York Times*, June 12.

54 Ballard, C. (2001) *Human Rights and the Mining Sector in Indonesia: A Baseline Study*. London: International Institute for Environment and Development, October.

55 Downing, T. E. (2002) *Avoiding New Poverty: Mining-Inducted Displacement and Resettlement*. London: International Institute for Environment and Development, April.

56 Bullard, R. D. (1996) Symposium: The legacy of American apartheid and environmental racism. *St. John's Journal of Legal Commentary*, 9, 445–74.

57 Bullard, R. D. (1994) *Unequal Protection: Environmental Justice and Communities of Color*. San Francisco, CA: Sierra Club Books. Healy,

N., Stephens, J. C., & Malin, S. A. (2019) Embodied energy injustices: Unveiling and politicizing the transboundary harms of fossil fuel extractivism and fossil fuel supply chains. *Energy Research & Social Science*, 48, 219–34.

58 Newman, D. K., & Day, D. (1975) *The American Energy Consumer.* Cambridge, MA: Ballinger.

59 Steger, T. (2007) *Making the Case for Environmental Justice in Central and Eastern Europe.* Budapest: CEU Center for Environmental Law and Policy.

60 Walker, W. (2012) *Environmental Justice: Concepts, Evidence, and Politics.* London: Routledge.

61 Blowers, A., & Leroy, P. (1994) Power, politics and environmental inequality: A theoretical and empirical analysis of the process of "peripheralization." *Environmental Politics*, 3(2), 197–228.

62 Rasmussen, C. M. (1998) Getting access to billions of dollars and having a nuclear waste backyard. *Journal of Land Resources and Environmental Law*, 18, 335–67.

63 Park, J., & Sovacool, B. K. (2018) The contested politics of the Asian atom: Peripheralisation and nuclear power in South Korea and Japan. *Environmental Politics*, 27(4), 686–711.

64 Ottaviano, D. M. (2003) *Environmental Justice: New Clean Air Act Regulations and the Anticipated Impact on Minority Communities.* New York: Lawyer's Committee for Civil Rights Under Law.

65 Swartz, A. (1994) Environmental justice: A survey of the ailments of environmental racism. *The Social Justice Law Review*, 2, 35–7.

66 Matsuo, T., & Schmidt, T. S. Managing tradeoffs in green industrial policies: The role of renewable energy policy design. *World Development*, 122, 11–26.

67 Yenneti, K., Day, R., & Golubchikov, O. (2016) Spatial justice and the land politics of renewables: Dispossessing vulnerable communities through solar energy mega-projects. *Geoforum*, 76, 90–9.

68 Temper, L. (2018) From boomerangs to minefields and catapults: Dynamics of trans-local resistance to land-grabs. *The Journal of Peasant Studies*, 46, 188–216.

69 Green, F. (2018) Anti-fossil fuel norms. *Climatic Change*, 150(1–2), 103–16.

70 Mangat, R., Dalby, S., & Paterson, M. (2017) Divestment discourse: War, justice, morality and money. *Environmental Politics*, 27(2), 187–208.

71 Buckley, T. (2019) Over 100 global financial institutions are exiting coal, with more to come. Available at http://ieefa.org/wp-content/uploads/2019/02/IEEFA-Report_100-and-counting_Coal-Exit_Feb-2019.pdf.

72 See https://follow-this.org/ and Neville, K. J., Cook, J., Baka, J., Bakker, K., & Weinthal, E. S. (2018) Can shareholder advocacy shape

energy governance? The case of the US antifracking movement. *Review of International Political Economy*, 26(1), 104–33.

73 Spreng, C., Sovacool, B. K., & Spreng, D. (2016) All hands on deck: Polycentric governance for climate change insurance. *Climatic Change*, 139(2), 129–40.

74 Van de Graaf, T., Haesebrouck, T., & Debaere, P. (2018) Fractured politics? The comparative regulation of shale gas in Europe. *Journal of European Public Policy*, 25(9), 1276–93. Blondeel, M., & Van de Graaf, T. (2018) Toward a global coal mining moratorium? A comparative analysis of coal mining policies in the USA, China, India and Australia. *Climatic Change*, 150(1–2), 89–101. Office of the Prime Minister (2016) United States-Canada Joint Arctic Leaders' Statement. Available at: https://pm.gc.ca/eng/news/2016/12/20/united-states-canada-joint-arctic-leaders-statement.

75 Setzer, J., & Vanhala, L. (2019) Climate change litigation: A review of research on courts and litigants in climate government. *WIREs Climate Change*, 10, e580.

76 Schwartz, J. (2018) New York sues Exxon Mobil, saying it deceived shareholders on climate change. *New York Times*, October 24.

77 Meckling, J., & Nahm, J. (2019) The politics of technology bans: Industrial policy competition and green goals for the auto industry. *Energy Policy*, 126, 470–9.

78 Birch, I., & Gilchrist, J. (2018) Survey of global activity to phase out internal combustion engine vehicles. Available at https://climateprotec tion.org/wp-content/uploads/2018/10/Survey-on-Global-Activities-to-Phase-Out-ICE-Vehicles-FINAL-Oct-3-2018.pdf.

79 See, e.g., Carrington, D. (2019) School climate strikes: 1.4 million people took part, say campaigners. *The Guardian*, March 19.

80 Subramanian, A. (2015) India is right to resist the West's carbon imperialism. Available at: https://www.ft.com/content/0805bac2-937d-11e5-bd82-c1fb87bef7af.

81 McGlade and Ekins, above, n. 36.

82 Kartha, S., Caney, S., Dubash, N. K., & Muttitt, G. (2018) Whose carbon is burnable? Equity considerations in the allocation of "right to extract." *Climatic Change*, 150(1–2), 117–29. Caney, S. (2016) *Climate Change, Equity, and Stranded Assets*. New York: Oxfam America.

83 Heffron & McCauley, above, n. 4. Healy & Barry, above, n. 4. Newell, P., & Mulvaney, D. (2013) The political economy of the "just transition." *The Geographical Journal*, 179(2), 132–40. Smith, S. (2017) *The Imperative of a Just Transition*. Brussels: Just Transition Center. Swilling, M., & Annecke, E. (2012) *Just Transitions: Explorations of Sustainability in an Unfair World*. Cape Town: UCT Press.

84 ILO (2018) *Just transition towards environmentally sustainable economies and societies for all*. ILO ACTRAV Policy Brief. Geneva: International Labour Organization.

85 Neslen, A. (2018) Spain to close most coalmines in €250m transition deal. *The Guardian*, October 26. Available at https://www.theguardian. com/environment/2018/oct/26/spain-to-close-most-coal-mines-after-striking-250m-deal.

86 Martello, M. L., & Jasanoff, S. (2004) Globalization and environmental governance. In Jasanoff, S., & Martello, M. L. (Eds.) *Earthly Politics: Local and Global in Environmental Governance*. Cambridge, MA: MIT Press, pp. 1–29.

Chapter 7   Energy technologies and innovation

1 IRENA (2019) *Renewable Power Generation Costs in 2018*. Abu Dhabi: IRENA.

2 IEA (2019) *Global Energy and CO$_2$ Status Report*. Paris: International Energy Agency.

3 Parag, Y., & Janda, K. B. (2014) More than filler: Middle actors and socio-technical change in the energy system from the "middle-out." *Energy Research & Social Science*, 3, 102–12. Jefferson, M. (2014) Closing the gap between energy research and modelling, the social sciences, and modern realities. *Energy Research & Social Science*, 4, 42–52. Di Lucia, L., & Ericsson, K. (2014) Low-carbon district heating in Sweden – Examining a successful energy transition. *Energy Research & Social Science*, 4, 10–20. Schubert, D. K. J., Thuß, S., & Möst, D. (2015) Does political and social feasibility matter in energy scenarios? *Energy Research & Social Science*, 7, 43–54.

4 Laird, F. N. (2013) Against transitions? Uncovering conflicts in changing energy systems. *Science as Culture*, 22(2), 149–56.

5 Hirsh, R. F., & Jones, C. F. (2014) History's contributions to energy research and policy. *Energy Research & Social Science*, 1, 106–11.

6 Araújo, K. (2014) The emerging field of energy transitions: Progress, challenges, and opportunities. *Energy Research & Social Science*, 1, 112–21.

7 Brown, M. A., & Sovacool, B. K. (2011) *Climate Change and Global Energy Security: Technology and Policy Options*. Cambridge, MA: MIT Press.

8 Similar categories of technologies have been suggested in previous reports. One typology is described in Kaya, Y. (1990) Impact of carbon dioxide emission control on GNP growth: Interpretation of proposed scenarios. Presentation, IPCC Energy and Industry Subgroup, Response Strategies Working Group, Paris. A second typology is offered in the 11-Lab Study: National Laboratory Directors (1997) *Technology Opportunities to Reduce US Greenhouse Gas Emissions*. Available at https://www.nrel.gov/docs/legosti/old/25680.pdf.

9 Schumpeter, J. (1942) *Capitalism, Socialism, Democracy*. New York: Harper and Brothers.

10 Tidd, J., & Bessant, J. R. (2018) *Managing Innovation: Integrating Technological, Market and Organizational Change*. Chichester: John Wiley & Sons.

11 Rogers, E. M. (1995) *Diffusion of Innovations*, 4th edn. New York: Free Press.

12 REN21 (2018) *Global Status Report*. Paris: REN21.

13 International Energy Agency (2019) *Electricity Information 2019: Overview*. Paris: OECD/IEA.

14 International Energy Agency (2018) *World Energy Outlook 2018*. Paris: IEA/OECD, p. 91.

15 Patt, A. (2015) *Transforming Energy: Solving Climate Change with Technology Policy*. Cambridge: Cambridge University Press.

16 Energy Transitions Commission (2018) Mission possible: Reaching net-zero carbon emissions from harder-to-abate sectors by mid-century. Available at http://www.energy-transitions.org/sites/default/files/ETC_MissionPossible_FullReport.pdf.

17 Lovins, A. B. (2007) Energy myth nine – Energy efficiency improvements have already reached their potential. In Sovacool, B. K. & Brown, M. A. (Eds.) *Energy and American Society – Thirteen Myths*. Dordrecht: Springer, p. 250.

18 International Energy Agency (2017) *Energy Efficiency Market Report 2017*. Paris: OECD/IEA.

19 Intergovernmental Panel on Climate Change (IPCC) (2007) *Climate Change 2007: Mitigation of Climate Change*. New York: Cambridge University Press, p. 325. International Energy Agency (2019) Transport. Available at https://www.iea.org/topics/transport/.

20 International Energy Agency (2019) CO_2 Emissions: Focus on Transport. Available at https://www.iea.org/statistics/co2emissions/.

21 International Energy Agency (2019) *Global EV Outlook 2019: Scaling Up the Transition to Electric Mobility*. Paris: OECD.

22 ERG. 2018. Clean Cobalt. ERG's Clean Cobalt Framework.

23 National Research Council of the National Academies (2009) *Transitions to Alternative Transportation Technologies – A Focus on Hydrogen*. Washington, DC: The National Academies Press.

24 International Energy Agency (2008) *World Energy Outlook 2008*, p. 506.

25 IPCC, above, n. 19, p. 389.

26 Intergovernmental Panel on Climate Change (IPCC) (2007) Working Group III: Fourth Assessment Report, chapter 6. Available at https://www.ipcc.ch/site/assets/uploads/2018/03/ar4_wg2_full_report.pdf.

27 Ibid., 11–12.

28 Xu, J. (2010) Local implementation of building energy policies in China's Jiangsu Province. In *Proceedings of the ACEEE Summer Study on Energy Efficiency in Buildings, August 17, 2010, Pacific Grove, CA, Volume 8*.

29 Energy Information Administration (2006) *Annual Energy Outlook 2006*, DOE/EIA-0383(2006). Washington, DC: US Department of Energy. EIA (2007) *Annual Energy Outlook 2007.*
30 Craford, M. G. (2008) High power LEDs for solid state lighting: Status, trends, and challenges. *Journal of Light and Visual Environment*, 32(2), 58–62.
31 Dunn, B., Kamath, H., & Tarascon, J.-M. (2011) Electrical energy storage for the grid: A battery of choices. *Science*, 334, 928–35. Simon, P., Gogotsi, Y., & Dunn, B. (2014) Where do batteries end and supercapacitors begin? *Science*, 343, 1210–11. Mwasilu, F., Justo, J. J., Kim, E.-Y., Do, T. D., & Jung, J.-W. (2014) Electric vehicles and smart grid interaction: A review on vehicle to grid and renewable energy sources integration. *Renewable and Sustainable Energy Reviews*, 34, 501–16. UK Parliamentary Office of Science and Technology (2015) Energy Storage Postnote, No. 492, April.
32 Sovacool, B. K. (2008) Valuing the greenhouse gas emissions from nuclear power: A critical survey. *Energy Policy*, 36(8), 2940–53.
33 Renewables Global Status Report: 2018 Update, REN21. Available at http://www.ren21.net/pdf/RE_GSR_2018_Update.pdf.
34 Ibid.
35 Ibid.
36 Sivaram, V. (2018) *Taming the Sun: Innovations to Harness Solar Energy and Power the Planet.* Cambridge, MA: MIT Press, p. xiv.
37 This Report – *The World Ocean Review* – can be downloaded at: http://worldoceanreview.com/en/wor-1/energy/.
38 This press release – "Potential of Renewable Energy Outlined in Report by the Intergovernmental Panel on Climate Change" – can be accessed via the IPCC website at: http://srren.ipcc-wg3.de/press/content/potential-of-renewable-energy-outlined-report-by-the-intergovernmental-panel-on-climate-change.
39 IRENA (2018) *Global Energy Transformation: A Roadmap to 2050.* Abu Dhabi: International Renewable Energy Agency.
40 Boyle, G. (2004) *Renewable Energy: A Power for a Sustainable Future*, 2nd edn. Oxford: Oxford University Press.
41 US Department of Energy (2018) Annual Capacity Factors for Electricity Generators in the United States. Washington, DC: DOE.
42 Childs, B., & Bradley, R. (2007) *Plants at the Pump: Biofuels, Climate Change, and Sustainability.* Washington, DC: World Resources Institute.
43 National Renewable Energy Laboratory (NREL) (n.d.) Learning about renewable energy – biofuels. Available at https://energyliteracy-principles.org/resource.aspx?id=582&title=NREL%20(National%20Renewable%20Energy%20Laboratory):%20Learning%20About%20Renewable%20Energy%20-%20Biofuels%20Basics.
44 DOE-EERE, *Renewable Energy Databook.* Available at http://www1.eere.energy.gov/maps_data/pdfs/eere_databook.pdf.

45 Shunmugam, V. (2009) Biofuels breaking the myth of "indestructible energy"? *Margin: The Journal of Applied Economic Research*, 3, 173–89; see also Johnson, F. X., & Virgin, I. (2010) Future trends in biomass resources for food and fuel. In Rosillo-Calle, F., & Johnson, F. X. (Eds.) *Food versus Fuel: An Informed Introduction to Biofuels*. London: Zed Books, pp. 164–90.

46 Sovacool, B. K., Brown, M. A., & Valentine, S. V. (2016) *Fact and Fiction in Global Energy Policy: Fifteen Contentious Questions*. Baltimore, MD: Johns Hopkins University Press.

47 Shunmugam, above, n. 45.

48 Farrell, A. E., Plevin, R. J., Turner, B. T., Jones, A. D., O'Hare, M., & Kammen, D. M. (2006) Ethanol can contribute to energy and environmental goals. *Science*, 311, 506.

49 National Research Council, above, n. 23, p. 1.

50 International Energy Agency (2019) *The Future of Hydrogen: Seizing Today's Opportunities*. Paris: OECD.

51 Deutch, J. M. et al. (2009) *Update of the MIT 2003 Future of Nuclear Power*. Cambridge, MA: MIT. Available at http://large.stanford.edu/courses/2017/ph241/kim-d2/docs/nuclearpower-update2009.pdf. National Commission on Energy Policy (2004) *Ending the Energy Stalemate*. Washington, DC: NCEP. Available at https://bipartisanpolicy.org/report/ending-energy-stalemate/

52 Fertel, M. S. (2010) The future of nuclear energy. Presentation to the National Academies Board on Energy and Environmental Systems, Washington, DC, September 14.

53 National Commission on Energy Policy, above, n. 51.

54 Deutch et al., above, n. 51, p. 4.

55 US Department of Energy (2006) *Carbon Sequestration Technology Roadmap and Program Plan 2006*. DOE, pp. 11–12. http://fossil.energy.gov/sequestration/publications/programplans/2006/2006_sequestration_roadmap.pdf (no longer available).

56 US Climate Change Technology Program (2003) *Technology Options for the Near and Long Term*. Washington, DC: CCTP.

57 US Department of Energy (n.d.) Terrestrial sequestration research. Available at http://www.fossil.energy.gov/programs/sequestration/terrestrial/index.html

58 Lewandrowski, J., Peters, M., Jones, C., House, R., Sperrow, M., Eve, M., & Paustian, K. (2004) Economics of sequestering carbon in the US agricultural sector. Technical Bulletin Number 1909. Washington, DC: US Department of Agriculture, Economic Research Service. US EPA (2005) Greenhouse gas mitigation potential in US forestry and agriculture. EPA 430-R-05-006. Washington, DC: EPA.

59 International Energy Agency and Nordic Energy Research (2013) *Nordic Energy Technology Perspectives: Pathways to a Carbon Neutral Energy Future*. Paris: OECD.

60 Sovacool et al., above, n. 46.
61 Brown and Sovacool, above, n. 7.
62 Unruh, G. (2000) Understanding carbon lock-in. *Energy Policy*, 28, 817–30.
63 Brown and Sovacool, above, n. 7.
64 Sovacool, B. K. (2017) Reviewing, reforming, and rethinking global energy subsidies: Towards a political economy research agenda. *Ecological Economics*, 135, 150–63.
65 Brown, M. A., & Chandler, S. (2008) Governing confusion: How statutes, fiscal policy, and regulations impede clean energy technologies. *Stanford Law and Policy Review*, 19(3), 472–509.
66 Sabety, T. (2005) Nanotechnology innovation and the patent thicket: Which IP policies promote growth? *Albany Law Journal of Science & Technology*, 15, 477–515.
67 Saunders, K. M., & Levine, L. (2004) Better, faster, cheaper-later: What happens when technologies are suppressed. *Michigan Telecommunications and Technology Law Review*, 11, 23–69.
68 Sovacool, B. K. (2008) Placing a glove on the invisible hand: How intellectual property rights may impede innovation in energy research and development (R&D). *Albany Law Journal of Science & Technology*, 18(2), 381–440, at p. 422.
69 Quinn, E. R. (2006) Cost of obtaining a patent. *Intellectual Property Watchdog*, http://www.ipwatchdog.com/patent_cost.html (no longer available).
70 Potts, J. R. What does it cost to obtain a patent, http://pw1.netcom.com/~patents2/What%20Does%20It%20Cost%20Patent.htm (no longer available).
71 Sovacool, above, n. 68, p. 429.
72 Biermayer, P., & Lin, J. (2004) Clothes washer standards in China – The problem of water and energy trade-offs in establishing efficiency standards. In *2004 ACEEE Summer Study on Energy Efficiency in Buildings*. Washington, DC: ACEEE.
73 Chappells, H., & Shove, E. (2004) *Comfort: A Review of Philosophies and Paradigms*. Lancaster: Lancaster University. Wilhite, H., Nakagami, H., Masuda, T., Yamaga, Y., & Haneda, H. (1996) A cross-cultural analysis of household energy use behavior in Japan and Norway. *Energy Policy*, 24(9), 795–803.
74 Sovacool, B. K. (2008) *The Dirty Energy Dilemma: What's Blocking Clean Power in the US*. Westport, CT: Praeger, p. 169.
75 Farhar, B. C. (1994) Trends in US public perceptions and preferences on energy and environmental policy. *Annual Review of Energy and Environment*, 9, 211–39.
76 Kentucky Environmental Education Council (2005) *The 2004 Survey of Kentuckians' Environmental Knowledge, Attitudes and Behaviors*. Frankfort, KY: KEEC.

77 Shelton, S. C. (2006) *The Consumer Pulse Survey on Energy Conservation.* Knoxville, TN: Shelton Group.

78 Hess, G. (2006) Bush promotes alternative fuel. *Chemical and Engineering News,* March 6, pp. 50–8.

79 Carrico, A. R., Padgett, P., Vandenbergh, M. P., Gilligan, J., & Wallston, K. A. (2009) Costly myths: An analysis of idling beliefs and behavior in personal motor vehicles. *Energy Policy,* 39, 2881–8.

80 Turrentine, T. S., & Kurani, K. S. (2007) Car buyers and fuel economy? *Energy Policy,* 35, 1213–23.

81 Rosoff, S. M., Pontell, H. N., & Tillman, R. (1998) *Profit Without Honor: White-Collar Crime and the Looting of America.* Upper Saddle River, NJ: Prentice Hall, p. viii.

82 Palm, J., & Lofstrom, E. (2008) Domestication of new technology in households. Presentation to the Society for the Social Studies of Science & European Association for the Study of Science and Technology Annual Conference, Rotterdam, The Netherlands, August 20–23.

83 Wallenborn, G. (2007) The new culture of energy: How to empower energy users? Proceedings of the International Conference on "Energy and Culture," February 7–8, Esbjerg, Denmark.

84 Sovacool, B. K., & Blyth, P. L. (2015) Energy and environmental attitudes in the green state of Denmark: Implications for energy democracy, low carbon transitions, and energy literacy. *Environmental Science & Policy,* 54, 304–15.

85 Feng, D., Sovacool, B. K., & Vu, K. M. (2010) The barriers to energy efficiency in China: Assessing household electricity savings and consumer behavior in Liaoning Province. *Energy Policy,* 38(2), 1202–9.

86 National Geographic/GlobeScan (n.d.) Consumer Greendex. Available at http://www.nationalgeographic.com/greendex/.

87 Prindle, B. et al. (2006) *Quantifying the Effects of Market Failures in the End-Use of Energy.* Paris: OECD/IEA.

88 Rezessy, S., Dimitrov, K., Urge-Vorsatz, D., & Baruch, S. (2006) Municipalities and energy efficiency in countries in transition: Review of factors that determine municipal involvement in the markets for energy services and energy efficient equipment, or how to augment the role of municipalities as market players. *Energy Policy,* 34(2), 223–37.

89 Brown, M. A., Chandler, J., Lapsa, M. V., & Sovacool, B. K. (2008) *Carbon Lock-In: Barriers to Deploying Climate Change Mitigation Technologies.* Oak Ridge, TN: Oak Ridge National Laboratory, p. 93.

90 D&R International (2008) *2008 Buildings Energy Data Book,* Table 2.2.2. Available at http://web.archive.org/web/20130214012611/http://buildingsdatabook.eren.doe.gov/DataBooks.aspx.

91 Ibid., Table 3.2.3.

92 Brown, M. A., Southworth, F., & Stovall, T. K. (2005) *Towards a Climate-friendly Built Environment.* Oak Ridge, TN: Oak Ridge National Laboratory

93 Smil, V. (2010) *Energy Transitions: History, Requirements, Prospects.* Santa Barbara, CA: Praegar, p. 150.
94 Knox-Hayes, J. (2012) Negotiating climate legislation: Policy path dependence and coalition stabilization. *Regulation & Governance*, 6(4), 545–67.
95 Goldthau, A., & Sovacool, B. K. (2012) The uniqueness of the energy security, justice, and governance problem. *Energy Policy*, 41, 232–40.
96 Meadowcroft, J. (2009) What about the politics? Sustainable development, transition management, and long term energy transitions. *Policy Sciences*, 42(4), 323–40.
97 Unruh, G. C. (2000) Understanding carbon lock-in. *Energy Policy*, 28, 817–30.
98 Geels, F. W., & Schot, J. W. (2007) Typology of sociotechnical transition pathways. *Research Policy*, 36, 399–417. Schot, J. W., & Geels, F. W. (2008) Strategic niche management and sustainable innovation journeys: Theory, findings, research agenda, and policy. *Technology Analysis and Strategic Management*, 20(5), 537–54.
99 Sovacool, B. K. (2016) How long will it take? Conceptualizing the temporal dynamics of energy transitions. *Energy Research & Social Science*, 13, 202–15.
100 Araujo, K. M. (2013) Energy at the frontier: Low carbon energy system transitions and innovation in four prime mover countries. PhD Dissertation, MIT.
101 IRENA (2018) *Global Energy Transformation: A Roadmap to 2050.* Abu Dhabi: International Renewable Energy Agency.
102 Ibid.
103 Ibid.
104 Ibid.
105 Ibid.
106 McKinsey & Company (2019) *Global Energy Perspective 2019: Reference Case.* Available at https://www.mckinsey.com/industries/oil-and-gas/our-insights/global-energy-perspective-2019.

Chapter 8   National and regional energy policy

1 International Energy Agency (2009) *World Energy Outlook 2009.* Paris: IEA/OECD, p. 41.
2 McGowan, F. (1996) Energy policy. In Kassim, H., & Menon, A. (Eds.) *The European Union and National Industrial Policy.* London: Routledge, pp. 132–52.
3 Hovi, J., Sprinz, D. F., & Underdal, A. (2009) Implementing long-term climate policy: Time inconsistency, domestic politics, international anarchy. *Global Environmental Politics*, 9(3), 20–39.
4 Tosun, J. (2018) Investigating ministry names for comparative policy analysis: Lessons from energy governance. *Journal of Comparative Policy Analysis: Research and Practice*, 20(3), 324–35.

5 Dubash, N. K. (2011) From norm taker to norm maker? Indian energy governance in global context. *Global Policy*, 2, 66–79.

6 Kong, B. (2011) Governing China's energy in the context of global governance. *Global Policy*, 2, 51–65.

7 McCormick, J. (2018) *Environmental Politics and Policy*. London: Macmillan, pp. 238–9.

8 Gunningham, N. (2013) Managing the energy trilemma: The case of Indonesia. *Energy Policy*, 54, 184–93.

9 Bradshaw, M. (2013) *Global Energy Dilemmas*. Cambridge: Polity.

10 Meckling, J., Kelsey, N., Biber, E., & Zysman, J. (2015) Winning coalitions for climate policy: Green industrial policy builds support for carbon regulation. *Science*, 349 (6253), 1170–1.

11 Newell, P., & Lane, R. (2018) A climate for change? The impacts of climate change on energy politics. *Cambridge Review of International Affairs*, https://doi.org/10.1080/09557571.2018.1508203.

12 Schmidt, T. S., & Sewerin, S. (2017) Technology as a driver of climate and energy politics. *Nature Energy*, 2(6), 17084.

13 Duffield, J. S. (2015) *Fuels Paradise: Seeking Energy Security in Europe, Japan, and the United States*. Baltimore, MD: Johns Hopkins University Press.

14 Breetz, H., Mildenberger, M., & Stokes, L. (2018) The political logics of clean energy transitions. *Business and Politics*, 20(4), 492–522. Meadowcroft, J. (2009) What about the politics? Sustainable development, transition management, and long term energy transitions. *Policy Sciences*, 42(4), 323–40.

15 Hall, P. A. (1997) The role of interests, institutions, and ideas in the comparative political economy of the industrialized nations. In Lichbach, M. I., & Zuckerman, A. S. (Eds.) *Comparative Politics: Rationality, Culture, and Structure*. Cambridge: Cambridge University Press, pp. 174–207.

16 Goldthau, A. (2012) From the state to the market and back: Policy implications of changing energy paradigms. *Global Policy*, 3(2), 198–210.

17 Altenburg, T., & Assmann, C. (2017) *Green Industrial Policy: Concept, Policies, Country Experiences*. Geneva, Bonn: UN Environment; German Development Institute/Deutsches Institut für Entwicklungspolitik (DIE).

18 Sovacool, B. K., & Mukherjee, I. (2011) Conceptualizing and measuring energy security: A synthesized approach. *Energy*, 36(8), 5343–55. Fischhendler, I., & Nathan, D. (2014) In the name of energy security: The struggle over the exportation of Israeli natural gas. *Energy Policy*, 70, 152–62.

19 Littlefield, S. R. (2013) Security, independence, and sustainability: Imprecise language and the manipulation of energy policy in the United States. *Energy Policy*, 52, 779–88.

20 Meckling, J. (2011) *Carbon Coalitions: Business, Climate Politics, and the Rise of Emissions Trading*. Cambridge, MA: MIT Press. Downie, C.

(2019) *Business Battles in the US Energy Sector: Lessons for a Clean Energy Transition*. New York: Routledge.

21 Newell, P., Jenner, N., & Baker, L. (2009) Governing clean development: A framework for analysis. *Development Policy Review*, 27(6), 717–39.

22 Ćetković, S., & Buzogány, A. (2016) Varieties of capitalism and clean energy transitions in the European Union: When renewable energy hits different economic logics. *Climate Policy*, 16(5), 642–57.

23 Mikler, J., & Harrison, N. E. (2012) Varieties of capitalism and technological innovation for climate change mitigation. *New Political Economy*, 17(2), 179–208.

24 Lachapelle, E., & Paterson, M. (2013) Drivers of national climate policy. *Climate Policy*, 13(5), 547–71.

25 Lockwood, M., Kuzemko, C., Mitchell, C., & Hoggett, R. (2017) Historical institutionalism and the politics of sustainable energy transitions: A research agenda. *Environment and Planning C: Politics and Space*, 35(2), 312–33.

26 Nixon, R. M. (1973) Address to the nation about policies to deal with the energy shortages. November 7. Available at www.presidency.ucsb.edu/ws/?pid=4034.

27 Herbstreuth, S. (2014) Constructing dependency: The United States and the problem of foreign oil. *Millennium – Journal of International Studies*, 43(1), 24–42.

28 Davis, D. H. (1993) *Energy Politics*. New York: St. Martin's Press.

29 Tomain, J. P. (2011) *Ending Dirty Energy Policy: Prelude to Climate Change*. Cambridge: Cambridge University Press.

30 Sovacool & Mukherjee, above, n. 18.

31 Laird, F. N. (2016) Avoiding transitions, layering change: The evolution of American energy policy. In Hager, C., & Stefes, C. H. (Eds.) *Germany's Energy Transition*. New York: Palgrave Macmillan, pp. 111–31.

32 Ibid.

33 Ibid., p. 119.

34 Ibid., p. 120.

35 Yao, L., & Chang, Y. (2014) Energy security in China: A quantitative analysis and policy implications. *Energy Policy*, 67, 595–604.

36 International Energy Agency (2010) *Energy Technology Perspectives 2010: Scenarios and Strategies to 2050*. Paris: IEA.

37 Yao, L., & Chang, Y. (2015) Shaping China's energy security: The impact of domestic reforms. *Energy Policy*, 77, 131–9.

38 Downs, E. (2004) The Chinese energy security debate. *The China Quarterly*, 177, 21–41.

39 Chang, T. P., & Hu, J. L. (2010) Total-factor energy productivity growth, technical progress, and efficiency change: An empirical study of China. *Applied Energy*, 87(10), 3262–70.

40 Jaffe, A. M., & Medlock III, K. B. (2005) China and Northeast Asia. In Kalicki, J. H., & Goldwyn, D. L. (Eds) *Energy and Security: Toward a New Foreign Policy Strategy*. Washington, DC: Woodrow Wilson Center Press, pp. 267–89. Janardhanan, N. (2010) Global energy geopolitics and the strategy of Chinese oil corporates. In Raghavan, V. R. (Ed.) *Emerging Challenges to Energy Security in the Asia Pacific*. Chennai: Centre for Security Analysis, pp. 70–86.

41 Yao & Chang, above, n. 35.

42 Ibid.

43 Yao & Chang, above, n. 37.

44 Geng, J.-B., & Ji, Q. (2014) Multi-perspective analysis of China's energy supply security, *Energy*, 64, 541–50.

45 Ibid. Sovacool, B. K. (2008) Placing a glove on the invisible hand: How intellectual property rights may impede innovation in energy research and development (R&D). *Albany Law Journal of Science & Technology*, 18(2), 381–440.

46 Lu, W., Su, M., Zhang, Y., Yang, Z., Chen, B., & Liu, G. (2014) Assessment of energy security in China based on ecological network analysis: A perspective from the security of crude oil supply. *Energy Policy*, 74, 406–13.

47 Lo, K. (2014) A critical review of China's rapidly developing renewable energy and energy efficiency policies. *Renewable and Sustainable Energy Reviews*, 29, 508–16.

48 Geng & Ji, above, n. 44.

49 Ren, J., & Sovacool, B. K. (2014) Enhancing China's energy security: Determining influential factors and effective strategic measures. *Energy Conversion and Management*, 88, 589–97. Ren, J., Andreasen, K. P., & Sovacool, B. K. (2014) Viability of hydrogen pathways that enhance energy security: A comparison of China and Denmark. *International Journal of Hydrogen Energy*, 39(28), 15320–9. Ren, J., & Sovacool, B. K. (2015) Prioritizing low-carbon energy sources to enhance China's energy security. *Energy Conversion and Management*, 92, 129–36.

50 Green, N., & Kryman, M. (2014) The political economy of China's energy and climate paradox. *Energy Research & Social Science*, 4, 135–8.

51 Own calculations based on the World Bank's World Development Indicators, available from: https://data.worldbank.org/.

52 UN DESA (2017) World Population Prospects: The 2017 Revision, United Nations Department of Economic and Social Affairs, Population Division.

53 International Energy Agency (2018) *World Energy Outlook 2018*, Paris: OECD/IEA.

54 Dietl, G. (2004) New threats to oil and gas in West Asia: Issues in India's energy security. *Strategic Analysis*, 28(3), 373–89.

55 Bambawale, M. J., & Sovacool, B. K. (2011) India's energy security: A sample of business, government, civil society, and university perspectives. *Energy Policy*, 39(3), 1254–64.

56 UNDP (2013) *Sustainable Energy for All: Country Brief India*. Bangkok: UNDP Asia Pacific Research Center.

57 Guay, J. (2013) *India, Coal Imports, and Energy Security*, Sierra Club. Available at http://sierraclub.typepad.com/compass/2013/07/india-coal-imports-energy-security.html.

58 UNDP, above, n. 56.

59 Ibid.

60 Ibid.

61 Ibid.

62 Jain, G. (2010) Energy security issues at household level in India. *Energy Policy*, 38(6), 2835–45.

63 Pachauri, S., & Jiang, L. (2008) The household energy transition in India and China. *Energy Policy*, 36, 4022–35.

64 UNDP, above, n. 56.

65 Ahn, S.-J., & Graczyk, D. (2012) *Understanding Energy Challenges in India: Policies, Players, and Issues*. Paris: International Energy Agency.

66 Ibid.

67 Sovacool, B. K., Imperiale, S., Gilbert, A., Eidsness, J., & Thomson, B. (2014) Troubled waters: The quest for electricity in water-constrained China, France, India, and the United States. *New York University Environmental Law Journal*, 21(3), 409–50.

68 Pode, R. (2010) Addressing India's energy security and options for decreasing energy dependency. *Renewable and Sustainable Energy Reviews*, 14(9), 3014–22.

69 Harris, G., & Bajaj, V. (2012) As power is restored in India, the "blame game" over blackouts heats up. *New York Times*, August 1.

70 International Energy Agency (2010) *Key World Energy Statistics*. Paris: OECD.

71 Ahn & Graczyk, above, n. 65.

72 Hashim, S. M. (2010) Power-loss or power-transition? Assessing the limits of using the energy sector in reviving Russia's geopolitical stature. *Communist and Post-Communist Studies*, 43(3), 263–74.

73 Dawson, J. I. (1996) *Eco-Nationalism: Anti-Nuclear Activism and National Identity in Russia, Lithuania, and Ukraine*. Durham, NC: Duke University Press, p. 174.

74 Marples, D. R. (1996) Nuclear politics in Soviet and Post-Soviet Europe. In Byrne, J., & Hoffman, S. M. (Eds.) *Governing the Atom: The Politics of Risk*. New Brunswick, NJ: Transaction Publishers, pp. 247–69.

75 BP (2019) *Statistical Review of World Energy*. Available at: www.bp.com/statisticalreview.

76 Movchan, A. (2015) Just an oil company? The true extent of Russia's dependency on oil and gas. Carnegie Moscow Center, September, 14. Available at https://carnegie.ru/commentary/61272.

77 Dalton, R. J., Garb, P., Lovrich, N. P., Pierce, J. C., & Whiteley, J. M. (1999) *Critical Masses: Citizens, Nuclear Weapons Production, and Environmental Destruction in the United States and Russia.* Cambridge, MA: MIT Press, p. 5.

78 Ibid., p. 370.

79 Hashim, above, n. 72.

80 Ibid.

81 Ibid.

82 Bilgin, M. (2011) Energy security and Russia's gas strategy: The symbiotic relationship between the state and firms. *Communist and Post-Communist Studies,* 44(2), 119–27.

83 Hashim, above, n. 72.

84 Ibid.

85 Mareš, M., & Laryš, M. (2012) Oil and natural gas in Russia's eastern energy strategy: Dream or reality? *Energy Policy,* 50, 436–48.

86 Kuleshov, D., Viljainen, S., Annala, S., & Gore, O. (2012) Russian electricity sector reform: Challenges to retail competition. *Utilities Policy,* 23, 40–9. Wengle, S. A. (2012) Post-Soviet developmentalism and the political economy of Russia's electricity sector liberalization. *Studies in Comparative International Development,* 47(1), 75–114. Pittman, R. (2007) Restructuring the Russian electricity sector: Re-creating California? *Energy Policy,* 35, 1872–83. Engoian, A. (2006) Industrial and institutional restructuring of the Russian electricity sector: Status and issues. *Energy Policy,* 34, 3233–44.

87 Cooke, D., Antonyuk, A., & Murray, I. (2012) *Toward a More Efficient and Innovative Electricity Sector in Russia.* Paris: International Energy Agency.

88 Bilgin, above, n. 82.

89 Cooke et al., above, n. 87.

90 Mareš & Laryš, above, n. 85.

91 Dunikov, D. O. (2015) Russia's view on development of novel and renewable energy sources, including hydrogen energy. *International Journal of Hydrogen Energy,* 40(4), 2062–3.

92 Boute, A., & Willems, P. (2012) RUSTEC: Greening Europe's energy supply by developing Russia's renewable energy potential, *Energy Policy,* 51, 618–29.

93 Baev, P. K. (2012) From European to Eurasian energy security: Russia needs and energy Perestroika. *Journal of Eurasian Studies,* 3(2), 177–84.

94 Goldemberg, J. (2013) *Brazil's Energy Story: Insights for US Energy Policy.* Washington, DC: The Aspen Institute.

95 Brown, M. A., & Sovacool, B. K. (2011) Brazil's Proalcohol program and promotion of flex-fuel vehicles. In *Climate Change and Global*

Energy Security: Technology and Policy Options. Cambridge, MA: MIT Press, pp. 260–74.

96 de Araújo, J. L. R. H., & de Oliveira, A. (2004) Brazilian energy policy: Changing course? *Oil, Gas & Energy Law Journal (OGEL),* 2(3).

97 Geller, H., Schaeffer, R., Szklo, A., & Tolmasquim, M. (2004) Policies for advancing energy efficiency and renewable energy use in Brazil. *Energy Policy,* 32(12), 1437–50.

98 de Araujo & de Oliveira, above, n. 96.

99 Kingstone, P. (2004) Critical Issues in Brazil's Energy Sector. Rice University, James A. Baker Institute for Public Policy, March.

100 Geller et al., above, n. 97.

101 Ibid.

102 Goldemberg, above, n. 94.

103 de Araujo & de Oliveira, above, n. 96.

104 Kingstone, above, n. 99.

105 Brown & Sovacool, above, n. 95.

106 de Freitas, L. C., & Kaneko, S. (2011) Decomposition of CO_2 emissions change from energy consumption in Brazil: Challenges and policy implications. *Energy Policy,* 39(3), 1495–504.

107 Kingstone, above, n. 99.

108 Goldemberg, above, n. 94.

109 Goldemberg, J., Schaeffer, R., Szklo, A., & Lucchesi, R. (2014) Oil and natural gas prospects in South America: Can the petroleum industry pave the way for renewables in Brazil? *Energy Policy,* 64, 58–70.

110 Pereira, M. G., Sena, J. A., Vasconcelos Freitas, M. A., & Fidelis da Silva, N. (2011) Evaluation of the impact of access to electricity: A comparative analysis of South Africa, China, India and Brazil. *Renewable and Sustainable Energy Reviews,* 15, 1427–41.

111 Hancock, K. J. (2015) The expanding horizon of renewable energy in sub-Saharan Africa: Leading research in the social sciences. *Energy Research & Social Science,* 5, 1–8.

112 Wabiri, N., & Amusa, H. (2010) Quantifying South Africa's crude oil import risk: A multi-criteria portfolio model. *Economic Modelling,* 27(1), 445–53.

113 Ibid.

114 Inglesi-Lotz, R., & Blignaut, J. N. (2011) South Africa's electricity consumption: A sectoral decomposition analysis. *Applied Energy,* 88(12), 4779–84.

115 Pereira et al., above, n. 110.

116 Matsika, R., Erasmus, B. F. N., & Twine, W. C. (2013) Double jeopardy: The dichotomy of fuelwood use in rural South Africa. *Energy Policy,* 52, 716–25.

117 World Bank Group (2010) Eskom investment support project questions & answers. April 14.

118 AfDB (2009) Project Appraisal Report: Medupi Power Plant, South Africa. October 14, p. 1.

119 Rafey, W., & Sovacool, B. K. (2011) Competing discourses of energy development: The implications of the Medupi coal-fired power plant in South Africa. *Global Environmental Change*, 21(3), 1141–51.

120 Pradhan, A., & Mbohwa, C. (2014) Development of biofuels in South Africa: Challenges and opportunities. *Renewable and Sustainable Energy Reviews*, 39, 1089–100.

121 International Energy Agency (2018) Electricity generation by fuel, South Africa 1990–2016. Available at https://www.iea.org/statistics/?country=SOUTHAFRIC.

122 Msimanga, B., & Sebitosi, A. B. (2014) South Africa's non-policy driven options for renewable energy development. *Renewable Energy*, 69, 420–7.

123 Ilskog, E., & Kjellstrom, B. (2008) And then they lived sustainably ever after? Assessment of rural electrification cases by means of indicators. *Energy Policy*, 36, 2674–84.

124 Madubansi, M., & Shackleton, C. M. (2005) Changing energy profiles and consumption patterns following electrification in five rural villages, South Africa. Department of Environmental Science, Rhodes University, South Africa.

125 Saadi, N., Miketa, A., & Howells, M. (2015) African clean energy corridor: Regional integration to promote renewable energy fueled growth. *Energy Research & Social Science*, 5, 130–2.

126 Green, N., Sovacool, B. L., & Hancock, K. (2015) Grand designs: Assessing the African energy security implications of the Grand Inga Dam. *African Studies Review*, 58(1), 133–58.

Chapter 9 Global energy governance

1 ElBaradei, M. (2008) A global agency is needed for the energy crisis. *Financial Times*, July 23.

2 Wilson, J. D. (2015) Multilateral organisations and the limits to international energy cooperation. *New Political Economy*, 20(1), 85–106.

3 McGowan, F. (2009) International regimes for energy: Finding the right level for policy. In Scrase, I., & MacKerron, G. (Eds.) *Energy for the Future: A New Agenda*. London: Palgrave Macmillan, pp. 20–34.

4 Florini, A., & Sovacool, B. K. (2009) Who governs energy? The challenges facing global energy governance. *Energy Policy*, 37(12), 5239–48.

5 Sovacool, B. K., & Florini, A. (2012) Examining the complications of global energy governance. *Journal of Energy & Natural Resources Law*, 30(3), 235–63.

6 Van de Graaf, T. (2013) *The Politics and Institutions of Global Energy Governance*. New York: Palgrave Macmillan.

7 Kaul, I., Conceicao, P., Le Goulven, K., & Mendoza, R. U. (Eds.) (2003) *Providing Global Public Goods: Managing Globalization*. Oxford: Oxford University Press.

8 Barrett, S. (2007) *Why Cooperate?: The Incentive to Supply Global Public Goods*. Oxford: Oxford University Press.

9 Goldthau, A. (2012) A public policy perspective on global energy security. *International Studies Perspectives*, 13(1), 65–84.

10 Karlsson-Vinkhuyzen, S. (2015) The legitimation of global energy governance: A normative exploration. In Mancebo, F., & Sachs, I. (Eds.) *Transitions to Sustainability*. Dordrecht: Springer, pp. 119–30.

11 Van de Graaf, T., & Colgan, J. (2016) Global energy governance: a review and research agenda. *Palgrave Communications*, 2(1), 15047. Dubash, N. K., & Florini, A. (2011) Mapping global energy governance. *Global Policy*, 2, 6–18.

12 Karlsson-Vinkhuyzen, S. I., Jollands, N., & Staudt, L. (2012) Global governance for sustainable energy: The contribution of a global public goods approach. *Ecological Economics*, 83, 11–18.

13 https://www.iea.org/topics/energysecurity/.

14 Elkind, J. (2010) Energy security: Call for a broader agenda. In Pascual, C., & Elking, J. (Eds.) *Energy Security: Economics, Politics, Strategies, and Implications*. Washington, DC: Brookings Institution Press, pp. 119–48.

15 Ruggie, J. (2011) Report of the special representative of the secretary-general on the issue of human rights and transnational corporations and other business enterprises: Guiding principles on business and human rights: implementing the United Nations "protect, respect and remedy" framework. Available at https://www.business-humanrights.org/sites/default/files/media/documents/ruggie/ruggie-guiding-principles-21-mar-2011.pdf.

16 Florini, A. (2008) Global governance and energy. In Pascual, C., & Elking, J. (Eds.) *Energy Security: Economics, Politics, Strategies, and Implications*. Washington, DC: Brookings Institution Press, pp. 149–81.

17 Raustiala, K., & Victor, D. G. (2004) The regime complex for plant genetic resources. *International Organization*, 58(2), 277–309. Colgan, J. D., Keohane, R. O., & Van de Graaf, T. (2012) Punctuated equilibrium in the energy regime complex. *The Review of International Organizations*, 7(2), 117–43.

18 Kérébel, C., & Keppler, J. H. (2009) *La gouvernance mondiale de l'énergie*. Paris: French Institute for International Relations. Sovacool, B. K., & Florini, A. (2012) Examining the complications of global energy governance. *Journal of Energy & Natural Resources Law*, 30(3), 235–63. Van de Graaf & Colgan, above, n. 11.

19 Hirst, N. (2018) *The Energy Conundrum: Climate Change, Global Prosperity, and the Tough Choices We All Have to Face*. London: World Scientific, p. 257.

20 Karlsson-Vinkhuyzen, S. I. (2010) The United Nations and global energy governance: Past challenges, future choices. *Global Change, Peace & Security*, 22(2), 175–95.

21 Lesage, D., Van de Graaf, T., & Westphal, K. (2010) *Global Energy Governance in a Multipolar World*. Farnham: Ashgate.

22 United Nations (2015) *Transforming our World: The 2030 Agenda for Sustainable Development*. Resolution adopted by the General Assembly. Available at https://sustainabledevelopment.un.org/post2015/trans-formingourworld.

23 OPEC (2018) Annual Statistical Bulletin. Available at https://asb.opec.org/index.php.

24 Morse, E. L. (1999) A new political economy of oil? *Journal of International Affairs*, 53(1), 1–29.

25 Yergin, D. (1991) *The Prize: The Epic Quest for Oil, Money, and Power*. New York: Simon & Schuster.

26 Colgan, J. D. (2014) The emperor has no clothes: The limits of OPEC in the global oil market. *International Organization*, 68(3), 599–632.

27 Ibid.

28 Van de Graaf, T. (2017) Is OPEC dead? Oil exporters, the Paris agreement and the transition to a post-carbon world. *Energy Research & Social Science*, 23, 182–8.

29 Yergin, above, n. 25.

30 Keohane, R. O. (1984) *After Hegemony: Cooperation and Discord in the World Political Economy*. Princeton, NJ: Princeton University Press, p. 223.

31 Mabro, R. (1991) *A Dialogue Between Oil Producers and Consumers: The Why and How*. Oxford: Oxford Institute for Energy Studies.

32 Katz, J. E. (1981) The International Energy Agency: Processes and prospects in an age of energy interdependence. *Studies in Comparative International Development*, 16(2), 67–85.

33 Van de Graaf, T. (2014) International Energy Agency. In Sperling, J. (Ed.) *Handbook of Governance and Security*. Cheltenham: Edward Elgar, pp. 489–503.

34 Scott, R. (1994) *The History of the International Energy Agency, 1974–1994 – Volume 1: Origins and Structure*. Paris: OECD/IEA.

35 Van de Graaf, T. (2012) Obsolete or resurgent? The International Energy Agency in a changing global landscape. *Energy Policy*, 48, 233–41.

36 Van de Graaf, T. (2015) The IEA, the new energy order and the future of global energy governance. In Lesage, D., & Van de Graaf, T. (Eds.) *Rising Powers and Multilateral Institutions*. London: Palgrave Macmillan, pp. 79–95.

37 Kissinger, H. A. (2009) The future role of the IEA. Speech on the Occasion of the 2009 Meeting of the IEA's Governing Board at the Ministerial Level, Paris, October 14.

38 Patrick, S., & Egel, N. (2016) The International Energy Agency's hybrid model. *Foreign Affairs*, August 9.

39 Van de Graaf, T. (2017) Organizational interactions in global energy governance. In Biermann, R., & Koops, J. A. (Eds.) *Palgrave Handbook of Inter-Organizational Relations in World Politics*. London: Palgrave Macmillan, pp. 591–609.

40 Fattouh, B., and van der Linde, C. (2011) *The International Energy Forum: Twenty Years of Producer – Consumer Dialogue in a Changing World*. Riyadh, Saudi Arabia: IEF, p. 58.

41 Griffin, J. (2010) Subroto: The Smiling Secretary General. Interview with Dr. Subroto. *OPEC Bulletin*, 3(10), 10–15.

42 Mitchell, J. (2005) Producer–consumer dialogue: What can energy ministers say to each other? Chatham House Report, November. London: Chatham. House.

43 Wagbara, O. (2007) How would the gas exporting countries forum influence gas trade? *Energy Policy*, 35(2), 1224–37.

44 Jaffe, A. M., & Soligo, R. (2006) Market structure in the new gas economy: Is cartelization possible? In Victor, D. G., Jaffe, A. M., & Hayes, M. H. (Eds.) *Natural Gas and Geopolitics from 1970 to 2040*. Cambridge: Cambridge University Press, pp. 439–64.

45 Clark, D. L. (1999) *A Citizen's Guide to the World Bank Inspection Panel*. Washington, DC: Center for International Environmental Law.

46 https://www.worldbank.org/en/who-we-are.

47 Mainhardt, H. (2019) World Bank Group financial flows undermine the Paris Climate Agreement. Available at https://urgewald.org/sites/default/files/World_Bank_Fossil_Projects_WEB.pdf.

48 Sovacool, B. K. (2017) Monitoring the moneylenders: Institutional accountability and environmental governance at the World Bank's Inspection Panel. *Extractive Industries & Society*, 4(4), 893–903.

49 Goldman, M. (2005) *Imperial Nature: The World Bank and Struggles for Social Justice in the Age of Globalization*. New Haven, CT: Yale University Press.

50 Tellam, I. (Ed.) (2000) *Fuel for Change: World Bank Energy Policy – Rhetoric vs Reality*. London: Zed Books.

51 Wood, J. R. (1993) India's Narmada river dams: Sardar Sarovar under siege. *Asian Survey*, 33(10), 968–84.

52 World Commission on Dams (2000) *Dams and Development: A New Framework for Decision-making. The Report of the World Commission on Dams*. London: Earthscan.

53 Friends of the Earth (2006) *How the World Bank's Energy Framework Sells the Climate and Poor People Short*. Washington, DC: Friends of the Earth.

54 Spalding-Fecher, R., Winkler, H., & Mwakasonda, S. (2005) Energy and the World Summit on Sustainable Development: What next? *Energy Policy*, 33(1), 99–112.

55 Keenan, J. H. (2005) Chad-Cameroon oil pipeline: World Bank & ExxonMobil in "Last Chance Saloon." *Review of African Political Economy*, 32, 395–405.

56 Pegg, S. (2009) Briefing: Chronicle of a death foretold: The collapse of the Chad-Cameroon pipeline project. *African Affairs*, 108(431), 311–20.

57 Mainhardt, above, n. 47.

58 Tienhaara, K., & Downie, C. (2018) Risky business? The Energy Charter Treaty, renewable energy, and investor-state disputes. *Global Governance*, 24(3), 451–71.

59 Hadfield, A., & Amkhan-Bayno, A. (2012) From Russia with cold feet: EU–Russia energy relations, and the Energy Charter Treaty. *International Journal of Energy Security and Environmental Research*, 1, 1–16.

60 Konoplyanik, A. (2017) The role of the ECT in EU–Russia energy relations. In Leal-Arcas, R., & Wouters, J. (Eds.) *Research Handbook on EU Energy Law and Policy*. Cheltenham: Edward Elgar, pp. 114–49.

61 Aalto, P. (2016) The new International Energy Charter: Instrumental or incremental progress in governance? *Energy Research & Social Science*, 11, 92–96.

62 Keay-Bright, S., & Defilla, S. (2019) Energy Charter Treaty review should end protection for fossil fuels. *Energy Post*, March 20. Available at https://energypost.eu/energy-charter-treaty-review-should-end-protection-for-fossil-fuels/.

63 Meyer, T. (2016) The World Trade Organization's role in global energy governance. In Van de Graaf, T. et al. (Eds.) *The Palgrave Handbook of the International Political Economy of Energy*. London: Palgrave Macmillan, pp. 139–71.

64 Desta, M. G. (2003) The Organization of Petroleum Exporting Countries, the World Trade Organization, and Regional Trade Agreements. *Journal of World Trade*, 37(3), 523–51.

65 Selivanova, Y. (2007) The WTO and energy: WTO rules and agreements of relevance to the energy sector. ICTSD Trade and Sustainable Energy Series Issue Paper No. 1. Geneva: International Centre for Trade and Sustainable Development.

66 UNCTAD (2000) *Trade Agreements, Petroleum and Energy Policies*. New York: United Nations, p. 15.

67 Ghosh, A. (2016) Clean energy trade conflicts: The political economy of a future energy system. In Van de Graaf, T. et al. (Eds.) *The Palgrave Handbook of the International Political Economy of Energy*. London: Palgrave Macmillan, pp. 175–204.

68 Lewis, J. I. (2014) The rise of renewable energy protectionism: Emerging trade conflicts and implications for low carbon development. *Global Environmental Politics*, 14(4), 10–35.

69 Rodrik, D. (2017) *Straight Talk on Trade: Ideas for a Sane World Economy*. Princeton, NJ: Princeton University Press, p. 259.

70 Downie, C. (2015) Global energy governance in the G-20: States, coalitions, and crises. *Global Governance*, 21(3), 475–492. Goldthau, A. (2017) The G20 must govern the shift to low-carbon energy. *Nature News*, 546(7657), 203. Andrews-Speed, P., & Shi, X. (2016) What role can the G20 play in global energy governance? Implications for China's Presidency. *Global Policy*, 7(2), 198–206.

71 Van de Graaf, T., & Westphal, K. (2011) The G8 and G20 as global steering committees for energy: Opportunities and constraints. *Global Policy*, 2, 19–30.

72 G20 Leaders' Statement (2009) Pittsburgh Summit, September 24–25. Available at http://www.g20.utoronto.ca/2009/2009communique0925. html.

73 Kirton, J. J., & Kokotsis, E. (2016) *The Global Governance of Climate Change: G7, G20, and UN Leadership*. London: Routledge.

74 Aldy, J. E. (2017) Policy surveillance in the G-20 fossil fuel subsidies agreement: Lessons for climate policy. *Climatic Change*, 144(1), 97–110.

75 Skovgaard, J. (2017) The devil lies in the definition: Competing approaches to fossil fuel subsidies at the IMF and the OECD. *International Environmental Agreements: Politics, Law and Economics*, 17(3), 341–53.

76 Victor, D. G. (2009) *The Politics of Fossil-Fuel Subsidies*. Geneva: IISD. Sovacool, B. K. (2017) Reviewing, reforming, and rethinking global energy subsidies: Towards a political economy research agenda. *Ecological Economics*, 135, 150–63. Van Asselt, H., & Skovgaard, J. (2018) *The Politics of Fossil Fuel Subsidies and Their Reform*. Cambridge: Cambridge University Press.

77 G20 (2014) G20 Energy Sustainability Working Group, Co-chairs' report, November 10. Available at https://g20.org/wp-content/uploads/2014/12/g20_energy_sustainability_working_group_2014_co-chairs_report.pdf.

78 IRENA (2018) Annual Report of the Director-General on the Implementation of the Work Programme and Budget for 2018–2019. A/9/3. December 12.

79 IRENA Statute (2009) Statute of IRENA signed in Bonn, Germany, on January 26, 2009.

80 Van de Graaf, T. (2013) Fragmentation in global energy governance: Explaining the creation of IRENA. *Global Environmental Politics*, 13(3), 14–33.

81 Steinbacher, K., & Pahle, M. (2016) Leadership and the Energiewende: German leadership by diffusion. *Global Environmental Politics*, 16(4), 70–89.

82 Andonova, L. B., Betsill, M. M., & Bulkeley, H. (2009) Transnational climate governance. *Global Environmental Politics*, 9(2), 52–73.

83 List available from: http://csonet.org/.

84 Pattberg, P., Biermann, F., Mert, A., & Chan, S. (Eds.) (2012) *Public-Private Partnerships for Sustainable Development: Emergence, Influence, and Legitimacy*. Cheltenham: Edward Elgar.

85 Sovacool, B. K., & Van de Graaf, T. (2018) Building or stumbling blocks? Assessing the performance of polycentric energy and climate governance networks. *Energy Policy*, 118, 317–24.

86 Abbott, K. W. (2012) Engaging the public and the private in global sustainability governance. *International Affairs*, 88(3), 543–64. Green, J. F. (2013) Order out of chaos: Public and private rules for managing carbon. *Global Environmental Politics*, 13(2), 1–25.

87 Van Alstine, J., & Andrews, N. (2016) Corporations, civil society, and disclosure: A case study of the extractive industries transparency initiative. In Van de Graaf, T. et al. (Eds) *The Palgrave Handbook of the International Political Economy of Energy*. London: Palgrave Macmillan, pp. 95–114.

88 Sovacool, B. K., Walter, G., Van de Graaf, T., & Andrews, N. (2016) Energy governance, transnational rules, and the resource curse: Exploring the effectiveness of the Extractive Industries Transparency Initiative (EITI). *World Development*, 83, 179–92.

89 De Coninck, H., Fischer, C., Newell, R. G., & Ueno, T. (2008) International technology-oriented agreements to address climate change. *Energy Policy*, 36(1), 335–56.

90 Szulecki, K., Pattberg, P., & Biermann, F. (2011) Explaining variation in the effectiveness of transnational energy partnerships. *Governance*, 24(4), 713–36.

91 Cherp, A., Jewell, J., & Goldthau, A. (2011) Governing global energy: Systems, transitions, complexity. *Global Policy*, 2(1), 75–88, at p. 76.

92 Van de Graaf, above, n. 6.

93 Colgan, J. D., Keohane, R. O., & Van de Graaf, T. (2012) Punctuated equilibrium in the energy regime complex. *The Review of International Organizations*, 7(2), 117–43.

94 Goldthau, A., & Sovacool, B. K. (2012) The uniqueness of the energy security, justice, and governance problem. *Energy Policy*, 41, 232–40.

95 Chan, S., van Asselt, H., Hale, T., et al. (2015) Reinvigorating international climate policy: A comprehensive framework for effective nonstate action. *Global Policy*, 6(4), 466–73. Hale, T. (2016) "All hands on deck": The Paris agreement and nonstate climate action. *Global Environmental Politics*, 16(3), 12–22.

96 Ostrom, E. (2010) Polycentric systems for coping with collective action and global environmental change. *Global Environmental Change*, 20(4), 550–7. Jordan, A. J., Huitema, D., Hildén, M., et al. (2015) Emergence of polycentric climate governance and its future prospects. *Nature Climate Change*, 5(11), 977–82.

97 Bäckstrand, K. (2008) Accountability of networked climate governance: The rise of transnational climate partnerships. *Global Environmental Politics*, 8(3), 74–12, at p. 78.

Chapter 10   Conclusions: Contested energy futures

1 Kander, A., Malanima, P., & Warde, P. (2015) *Power to the People: Energy in Europe over the Last Five Centuries*. Princeton, NJ: Princeton University Press.

2 UNCTAD (2019) Merchant fleet by flag of registration and by type of ship, annual, 1980–2018. Available at https://unctadstat.unctad.org/wds/TableViewer/tableView.aspx?ReportId=93.

3 UNCTAD (2018) *Review of Maritime Transport 2018*. Geneva: UNCTAD, p. 5.

4 Overland, I. (2016) Energy: The missing link in globalization. *Energy Research & Social Science*, 14, 122–30.

5 https://www.tennet.eu/our-grid/international-connections/norned/.

6 Parenteau, P., & Barnes, A. (2013) A bridge too far: Building off-ramps on the shale gas superhighway. *Idaho Law Review*, 49, 325.

7 Chazan, G. (2011) Japan to use more LNG. *Wall Street Journal*, March 16, p. 25.

8 Stafford, P., Blas, J., & Farchy, J. (2011) Nuclear problems put energy markets in a spin. *Financial Times*, March 17, p. 15.

9 Sovacool, B. K., & Cooper, C. J. (2013) *The Governance of Energy Megaprojects: Politics, Hubris, and Energy Security*. London: Edward Elgar.

10 Sharples, M. et al. (2006) Post Mortem Failure Assessment of MODUs During Hurricane Ivan. US Government Minerals Management Service.

11 Meckling, J., & Hughes, L. (2018) Global interdependence in clean energy transitions. *Business and Politics*, 20(4), 467–91.

12 Fraunhofer Institute for Solar Energy Systems (2019) Photovoltaics report. Available at https://www.ise.fraunhofer.de/content/dam/ise/de/documents/publications/studies/Photovoltaics-Report.pdf.

13 Jewell, J., Vetier, M., & Garcia-Cabrera, D. (2019) The international technological nuclear cooperation landscape: A new dataset and network analysis. *Energy Policy*, 128, 838–52.

14 Elbein, S. (2019) Europe's renewable energy policy is built on burning American trees. *Vox*, March 4.

15 Sovacool, B. K., Perea, M. A. M., Matamoros, A. V., & Enevoldsen, P. (2016) Valuing the externalities of wind energy: Assessing the environmental profit and loss of wind turbines in Northern Europe. *Wind Energy*, 19(9), 1623–47.

16 Dale, D. (2015) New economics of oil. Paper presented to Society of Business Economists Annual Conference, London. Available at https://www.bp.com/content/dam/bp/business-sites/en/global/corporate/pdfs/news-and-insights/speeches/new-economics-of-oil-spencer-dale.pdf.

17 United States Environmental Protection Agency (1997) Mercury Study Report to Congress, Volume I: Executive Summary, EPA-452/R-97-003.

18 Hopkin, M. (2005) Acid rain still hurting Canada. *Nature*, August 10.
19 Myers, S. L. (2010) Lament for a once-lovely waterway. *New York Times*, June 12; Myers, S. L. (2010) Vital river is withering, and Iraq has no answer. *New York Times*, June 12.
20 Ewing, J. J., & McRae, E. (2012) Transboundary haze in Southeast Asia: Challenges and pathways forward. NTS Alert, October.
21 Sovacool, B. K. (2011) *Contesting the Future of Nuclear Power*. Singapore: World Scientific Publishing.
22 Martin, C. (2013) Tea Party's Green faction fights for solar in Red States. Bloomberg Market News, November 12. Available at http://www.bloomberg.com/news/2013-11-12/tea-party-s-green-faction-fights-for-solar-in-red-states.html
23 Klinke, A., & Ortwin, R. (2006) Systemic risks as challenge for policy making in risk governance. *Forum: Qualitative Social Research*, 7(1), Art. 33, at p. 3. Sidortsov, R. (2014) Reinventing rules for environmental risk governance in the energy sector. *Energy Research & Social Science*, 1, 171–82.
24 Green, J., Hale, T., & Colgan, J. (2019) The existential politics of climate change. *Global Policy*, February 21. Available at https://www.globalpolicyjournal.com/blog/21/02/2019/existential-politics-climate-change.
25 Mercure, J. F., Pollitt, H., Viñuales, J. E., Edwards, N. R., Holden, P. B., Chewpreecha, U., ... & Knobloch, F. (2018) Macroeconomic impact of stranded fossil fuel assets. *Nature Climate Change*, 8(7), 588–93.
26 Burrow, S. (2017) Climate: Towards a just transition, with no stranded workers and no stranded communities. OECD Insights, May. Available at http://oecdinsights.org/2017/05/23/climate-towards-a-just-transition-with-no-stranded-workers-and-no-stranded-communities/.
27 Manley, D., Cust, J., & Cecchinato, G. (2017) Stranded nations: The climate policy implications for fossil fuel-rich developing countries. OxCarre Policy Paper, 34.

Index